Campden & Chorleywood Food
Research Association Group

Key Topics in Food
Science and Technology – No. 11

Raw materials and ingredients in food processing

Tim Hutton

Campden & Chorleywood Food Research Association Group comprises
Campden & Chorleywood Food Research Association
and its subsidiary companies
CCFRA Technology Ltd   CCFRA Group Services Ltd   Campden & Chorleywood Magyarország

© CCFRA 2005

Campden & Chorleywood Food
Research Association Group

Chipping Campden, Gloucestershire, GL55 6LD UK
Tel: +44 (0) 1386 842000   Fax: +44 (0) 1386 842100
www.campden.co.uk

Information emanating from this company is given after the exercise of all reasonable care and skill in its compilation, preparation and issue, but is provided without liability in its application and use.

The information contained in this publication must not be reproduced without permission from the CCFRA Publications Manager.

Legislation changes frequently. It is essential to confirm that legislation cited in this publication and current at the time of printing is still in force before acting upon it. Any mention of specific products, companies or trademarks is for illustrative purposes only and does not imply endorsement by CCFRA.

© CCFRA 2005
ISBN: 0 905942 74 4
A catalogue record for this book is available from the British Library.

SERIES PREFACE

Food and food production have never had a higher profile, with food-related issues featuring in newspapers or on TV and radio almost every day. At the same time, educational opportunities related to food have never been greater. Food technology is taught in schools, as a subject in its own right, and there is a variety of food-related courses in colleges and universities - from food science and technology through nutrition and dietetics to catering and hospitality management.

Despite this attention, there is widespread misunderstanding of food - about what it is, about where it comes from, about how it is produced, and about its role in our lives. One reason for this, perhaps, is that the consumer has become distanced from the food production system as it has become much more sophisticated in response to the developing market for choice and convenience. Whilst other initiatives are addressing the issue of consumer awareness, feedback from the food industry itself and from the educational sector has highlighted the need for short focused overviews of specific aspects of food science and technology with an emphasis on industrial relevance.

The *Key Topics in Food Science and Technology* series of short books therefore sets out to describe some fundamentals of food and food production and, in addressing a specific topic, each issue emphasises the principles and illustrates their application through industrial examples. Although aimed primarily at food industry recruits and trainees, the series will also be of interest to those interested in a career in the food industry, food science and technology students, food technology teachers, trainee enforcement officers and established personnel within industry seeking a broad overview of particular topics.

Leighton Jones
Series Editor

PREFACE TO THIS VOLUME

The chemical and physical characteristics of food raw materials and ingredients have a profound effect on the finished food product as consumed. In many cases individual starting materials have been developed with a particular type of processing condition and end product characteristic in mind. In some cases, whole industries have developed to tailor a particular raw material to a variety of end uses. In others, a range of ingredients have been developed to improve on the characteristics of the finished product.

This short book sets out to illustrate to those without any specific expertise in food processing, the many aspects of matching ingoing raw materials and ingredients to final product quality. This book has a strong industrial slant, and uses examples from food manufacturing to illustrate the many problems that the food industry has to face in producing food of the desired quality, and the solutions it has found to these problems. However, each individual ingredient, raw material and product has its own set of characteristics, so this book can only give illustrative examples. It looks at the characteristics of three of the world's main staples – wheat, rice and potatoes, before discussing the main characteristics and uses of the major ingredients in processed foods. It finishes by looking at some particular end products, such as meringues, ice cream and mayonnaise, to see how the combinations of raw ingredients interact, and how similar tasting end-products can be produced from different sets of ingredients.

Tim Hutton, CCFRA

ACKNOWLEDGEMENTS

Thanks as always are due to the many colleagues at CCFRA whose work I have referred to in producing this book, in particular to Sam Millar, Richard Stanley, Sarah Chapman and Leighton Jones for their technical expertise.

NOTE

All definitions, legislation, codes of practice and guidelines mentioned in this publication are included for the purpose of illustration only and relate to UK practice unless otherwise stated.

CONTENTS

1. Introduction 1

2. Properties of selected raw materials 3
   2.1 Rice 3
   2.2 Potatoes 10
   2.3 Wheat 17
   2.4 Conclusions 31

3. Specific ingredients 34
   3.1 Fats and fat substitutes 34
   3.2 Emulsifiers and stabilisers 39
   3.3 Starch and related thickeners 41
   3.4 Sugar 45
   3.5 Salt 46
   3.6 Gelling agents and thickeners 52
   3.7 Proteins 61
   3.8 Water 63
   3.9 Air 66
   3.10 Microbial cultures 68
   3.11 Other 'additives' 70

4. Examples of the use of ingredients for a specific functional purpose 74
   4.1 Laminated fat products 74
   4.2 Meringues 76
   4.3 Soufflés 77
   4.4 Ice cream 77
   4.5 Mayonnaise 80
   4.6 Texturised vegetable protein (TVP) 82

5. Conclusions 85

6. References 86

   About CCFRA 92

1. INTRODUCTION

Even before mankind progressed from being a hunter-gatherer to having a farming lifestyle, raw food materials could be blended together to add some culinary benefit to the final product, and improve or increase what may have been a limited diet. One of the first applications of ingredient functionality (i.e. the use of raw materials to give the food a specific physical characteristic) was probably the use of starchy roots and tubers to thicken stew-type meals. The physicochemical properties of starch means that it is still probably the most widely used functional ingredient in food recipes - either as the purified material in many industrially manufactured products, or as a component of raw materials such as potatoes, rice and flour. This has gone hand-in-hand with it being the most important nutritional ingredient in terms of energy for the body; most civilisations around the world have one or more starch-based ingredients as a staple food.

Early farmers bred both livestock and crops, mainly to increase yield and resistance to illness or disease. With livestock these factors are still the main driving forces behind breeding regimes, but with crops there has been an additional motivation - to produce raw materials with specific characteristics that enhance their functionality in processed products. Major examples are wheat and other cereals, and root vegetables such as potatoes. Wheat with higher protein levels can be processed to flour which is more suitable for breadmaking, and potatoes can now be found that are labelled as suitable for baking, roasting, boiling, chipping and various combinations of these.

The modern food industry has expanded on these basics of raw material functionality and uses a wide range of specific ingredients for various purposes in formulated foods. In some cases, these are 'whole' foods, such as eggs, which contain chemical components such as lecithins and albumens; the former are emulsifiers, and allow fat-soluble and water-soluble food components to be successfully mixed, for example in cakes; the latter are proteins which can be whisked to a foam, for example in the production of meringues. Alternatively, lecithins can be extracted from eggs or soya, for example, and used as 'additives': their role and functionality, however, is the same.

Fats and oils are widely used functional ingredients. Without their use, many food products would not exist: e.g. pies, biscuits, pastries. One of their major roles is in the development of the desired texture in the product, and what the sensory scientists describe as 'mouthfeel'. Sugar also has a significant influence on these properties.

So what is functionality in the context of this Key Topic? With primary products such as potatoes it is fairly easy to define: it is the fitness for the end use of the material. Are they a small, waxy variety that is suitable for canning, or are they large and floury and therefore more suited to oven baking? With isolated ingredients used in compound, 'formulated' foods, the scope becomes a little wider: it is the role that the ingredients play in producing and maintaining the structure and texture of the final product, and even the perceived flavour. Whilst this book will not look specifically at flavour compounds, it will address the issue of flavour as affected by major ingredients. Some ingredients, particularly salt and sugar, also have a major functional role in food preservation. This is broadly covered in the Key Topic on food preservation (Hutton, 2004), and will only be mentioned briefly here.

It has become common practice in the media to label ingredients as 'natural' or 'artificial'. Hopefully, this Key Topic will demonstrate that this supposedly clear distinction is often a little blurred at the edges, and that there is a direct link between whole ingredients and components that would be considered as 'additives'. The book is divided into three sections: raw materials, ingredients and specific products. In some cases, the assigning of a specific example to one section or another is somewhat arbitrary, as the interconnection of raw materials, ingredients and final product is seamless. Cross-references are given throughout the text to illustrate this.

2. PROPERTIES OF SELECTED RAW MATERIALS

Farmers and growers are now highly focused on the end uses of the crops that they are growing. Are the vegetables that are being grown destined for the 'fresh' or 'minimally processed' market, or will they be canned or frozen? Are the potatoes 'new boilers' or for baking, or are they destined to be oven-ready chips? Is the wheat going to be milled principally for bread flour, or is it a biscuit variety? Obviously, the farmer needs to know this from a commercial point of view, so that the correct decisions can be made as to what varieties to buy, and to whom to try and sell the harvested produce. Often, the destiny of the material will be stipulated as part of a contract and specification between the supplier and the customer before the crop is even sown. This chapter will look at some typical major examples of what factors influence the end-use suitability of these raw materials, and include some less familiar concepts that illustrate how importantly end-use functionality is now perceived. Three major food crops have been chosen - wheat, potatoes and rice - to illustrate these factors. Along the way other examples - vegetables for canning, varieties of apple and orange, milk in cheese production - are also used to provide further illustrations.

2.1 Rice

Rice is a cereal that has been developed for many different food product applications - both savoury and sweet. In the United States, which both produces and consumes a large quantity of rice, the normal milled white varieties are classified as long-grain, medium-grain or short-grain, depending on the length/width ratios, and a Standard exists for this classification (USDA, 1989). Among the many retail outlets for rice are as a milled raw product for home cooking as a savoury or sweet dish, as a frozen product, as a breakfast cereal and in snack bar type products, in canned puddings, and in a variety of savoury processed foods such as soups. It is also used as a starting material in the brewing industry. Different varieties with different structural characteristics and chemical compositions are bred to meet the demanding requirements of each of these end uses. The characteristics and requirements of some of these products are briefly reviewed below as an indication of the variety of alternatives available.

The first major point to consider is why rice is commonly milled to produce the familiar white product that is widely used. For any rice to be edible, it has to be hydrated (by cooking in a liquid) and the structure of the starch altered into a digestible form. (Cooking also allows this to happen - the nutritional aspects of starch are mentioned in a related Key Topic - see Hutton, 2002b). In all cases, the outer hull is removed to yield brown rice. This can then be milled to yield white rice. The bran layers that are removed by the milling process contain relatively higher levels of vitamins and minerals than the rest of the grain, and there has recently been an increase in the use of brown rice as a savoury accompaniment. From a processing point of view, however, it has quite different characteristics, due to the wax content in the outer brown layer (Wilkinson and Champagne, 2004). This reduces the rate of water absorption, increasing the length of cooking time from 15-25 minutes (for white rice) to 45-60 minutes. The texture of the cooked product is also different, being somewhat chewy.

The processing characteristics of the various rice varieties are largely related to the properties of the starch within the rice grain. Raw rice contains about 75-85% starch and so clearly the nature of the starch will influence the eating characteristics of the cooked product. The two main factors are the level of amylose in the starch and the gelatinisation temperature of the starch. Starch is not a 'single' molecule, but is made up of two distinct components: amylose and amylopectin. Amylose is a long, virtually unbranched chain of glucose molecules linked through the carbon atoms at positions 1 and 4. Amylopectin is a highly branched polymer with 15-30 of the (1-4) linkages in each branch, the branches being joined by linkages between the carbon atoms at positions 1 and 6 (see Figures 1 and 2)

Long grain rice varieties have amylose contents of 21-23%, whereas short- and medium-grain varieties have levels of 15-20%. Starch gelatinisation occurs at 69-73°C in the former and 61-63°C in the latter (see Box 1)

So what characteristics are required for different applications and how do these relate to the starch content?

Figure 1 - Molecular structure of glucose

The structure of monosaccharides can be depicted in several ways, but it should be remembered that in reality they are 3-dimensional. In this depiction of glucose ($C_6H_{12}O_6$), the standard numbering system for the carbon atoms is shown (see later for explanation on the structure of polysaccharides in general and starch in particular).

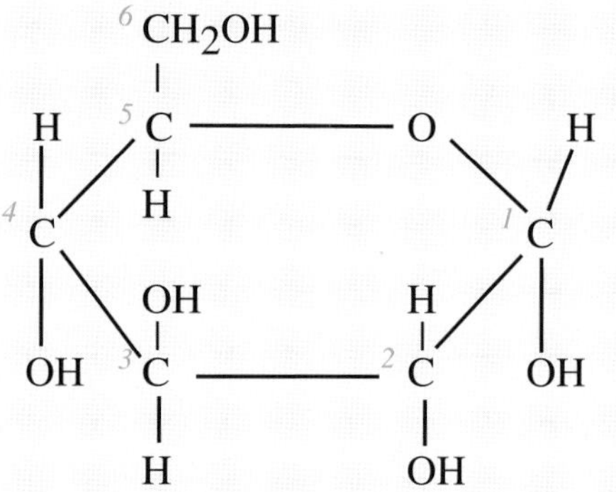

Figure 2 - Schematic diagram of amylose and amylopectin

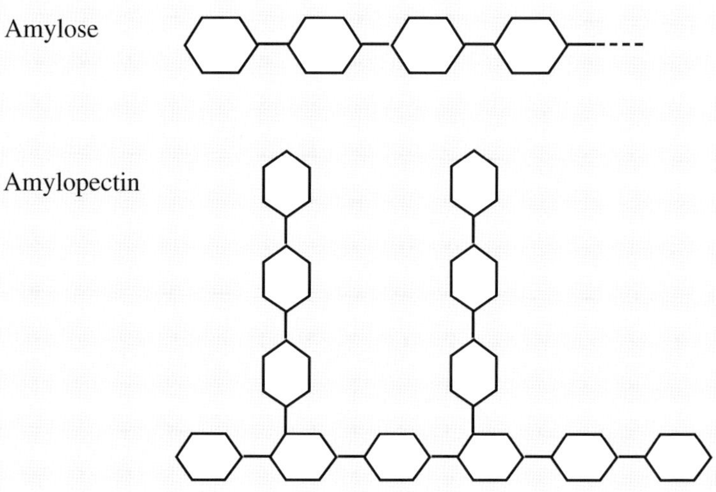

Amylose

Amylopectin

Box 1 - Starch gelatinisation and retrogradation

Starch is a highly complex and variable molecule. As well as broadly consisting of amylopectin and amylose fractions, these fractions themselves are heterogeneous, and different combinations of them join together with other minor components (including small amounts of protein and lipids) to form the starch granule. Some parts of the granule contain starch polymers that are crystalline in nature, while others are amorphous in structure. The phenomenon of gelatisation is exceedingly complex - it is the end result of a number of molecular structural changes in the starch granule. However, in broad terms, a smaller starch granule size and a higher amylose content both tend to increase gelatinisation temperature.

In gelatinisation, two fundamental changes occur: the degree of crystallinity is lost and the starch granule swells. These two phenomena happen together in some cases (e.g. in potato starch), whereas in others (e.g. wheat starch), they occur separately (in heat-induced gelatinisation swelling starts at about 20°C higher than the observed loss of crystallinity). The changes are the result of the breaking of the hydrogen bonds that hold the glucan chains of the individual amylose and amylopectin fractions together. This is usually achieved by heating the starch in water or milk (although other chemical treatments can achieve the same result), and can also be caused by milling - which is used in rice processing. Reversible (pre-gelatinisation) swelling can also occur from heating starch in water.

Retrogradation of starch is a phenomenon caused when linear amylose chains of the starch molecule reassociate via the formation of hydrogen bonds. These effectively increase the size of the molecule, and eventually this leads to the formation of insoluble particles. Retrogradation is associated with a hardening or drying of the rice grain, and with staling phenomena in other products (e.g. bread).

References:

Banks, W. and Greenwood, C.T. (1975) Starch and its components. Publ: Edinburgh University Press

Galliard, T. (Ed) (1987). Starch: Properties and Potential. Critical Reports on Applied Chemistry, Vol. 13. John Wiley

Box 2 - Quick cooking rice

Brown and milled white rice take approximately 45-60 and 15-25 minutes, respectively, to cook. In both domestic and catering outlets the pressure to reduce this led to the development of quick-cooking products. These are generally long-grain rice varieties that can be cooked in about 5 minutes. Many variations on the process to achieve this have been patented, but those generally used involve the raw, milled rice being pre-cooked and then dried so that the grains retain a porous and open structure without clumping. Clearly, the selection of cooking characteristics which allow a free-flowing product to be manufactured are of great importance. In other methods, dry heat and kernel abrasion are used, which do not gelatinise the starch.

Box 3 - Basmati rice

Basmati is the name given to a collection of aromatic rices grown in the Himalayan foothill regions of India and Pakistan. These have become very popular throughout the world, and command a significant price premium. These varieties are characterised by having long thin grains that approximately double in length on cooking, but with little increase in breadth, to yield a dagger-like shape. Although with a fluffy, soft texture when cooked, the grains are relatively low in amylose content with medium-low gelatinisation temperature (Bhattacharjee *et al*, 2002). In this respect they differ from aromatic varieties grown in the USA, for example, which have relatively high amylose contents and an intermediate starch gelatinisation temperature.

As well as varietal differences, the characteristics of Basmati rice varieties owe much to the climate and agricultural conditions in the regions in which they are grown. As a result, the term Basmati can only be applied to these special aromatic varieties grown in India and Pakistan.

References:

Bhattacharjee, P., Singhal, R.S., and Kulkarni, P.R. (2002) Basmati rice: a review. International Journal of Food Science and Technology, **37** (1): 1-12

Raw materials and ingredients in food processing

When choosing a rice variety for boiling as a savoury staple, one of the main characteristics required is that the cooked grains be able to flow freely and not clump together. Long-grain varieties generally provide this type of characteristic, and cultivars with amylose contents greater than 24% and increased resistance to overcooking have been developed. In the USA, one particular variety, Newrex, was developed to satisfy the needs of the food industry for a rice that could withstand the rigours of canning or the processing required for subsequent quick-cooking (see Box 2). Newrex typically has an amylose content of 28%.

The short and medium-grain varieties tend to have properties that make them more suitable for ready-to-eat breakfast cereals (e.g. rice crispy-type products - see below). They are also preferred for baby foods and other foods that are going to be served cold, because there is less retrogradation of the starch on cooling compared with long-grain rice (see Box 1).

A major use of medium- and short-grain varieties of rice is for breakfast cereals, especially those produced by puffing. There are several methods for puffing of cereals in general: in gun puffing and oven puffing, whole grains are used; in extrusion puffing, flour dough is used. Most cereals can be puffed via high-pressure extrusion from flour; gun puffing tends to be limited to rice and wheat, and oven puffing to rice and corn (maize). Crisped rice can be produced by either oven puffing or high-pressure extrusion puffing (Wilkinson and Champagne, 2004). Fast and Caldwell (2000) describe these and a range of other breakfast cereal manufacturing processes. With gun-puffed rice, the only significant pretreatment that the rice undergoes is normal milling to reduce the fat content to 0.5-1.5%. The grain is then cooked and subsequently subjected to a sudden pressure drop in the surrounding atmosphere: this causes the grain to puff. In oven puffing, the final product is achieved by subjecting the cooked and partially rolled and dried grains (rolling creates fissures in the grain) to very high temperatures - typically 290-340°C. The fissures help promote the puffing of the rice as the moisture suddenly vaporises and expands.

Arborio rice

These medium-grain varieties, which are typically used in risottos, have a large, bold kernel with a characteristic chalky centre (Wilkinson and Champagne, 2004).

Properties of raw materials

On cooking, they develop a creamy texture (through the release of starch) around a chewy centre and are particularly good at absorbing flavours. There is less need for the free-flowing nature of the individual grains found in longer-grain varieties.

Pudding rice

Varieties used for rice puddings are waxy, short-grain varieties, which are low in amylose and high in amylopectin. The individual grains tend to lose their shape when cooked, and become very sticky. These varieties are also used for a variety of other sweet and dessert applications, especially in Japan and other Oriental countries. These also utilise the sticky, soft nature of the cooked product, as well as its slower rate of retrogradation or hardening relative to that of cooked non-waxy rice.

Box 4 - Wild rice

Rather confusingly, wild rice is a term used to describe both uncultivated relatives of cultivated rice (*Oryza sativa* and other *Oryza* species) and the North American plants of the genus *Zizania*. The latter were regularly harvested from the wild and used as staple carbohydrate sources by several tribes of native North Americans. Although these original users did start to cultivate wild rice, by seeding suitable lakes and rivers, it is only since about 1950 that there has been an increase in the popularity of the product and a process of domestication and commercialisation.

Although wild rice had a niche market for some time due to its relatively good nutritional content (especially total protein and the amino acids lysine, threonine and methionine), it required special care in harvesting due to the shattering nature of the seeds, which necessitated several harvestings over a two- to three-week period. The selection of shattering-resistant cultivars and their cultivation greatly improved the availability of wild rice. The product is now increasingly used as a savoury accompaniment in place of traditional rice, and other uses (e.g. in soups and in blends with white, long-grain rice) are also growing.

References:

Oelke, E.A., Porter, R.A., Grombacher, A.W. and Addis, P.B. (1997) Wild rice - new interest in an old crop. Cereal Foods World, **42** (4), 234-247.

Raw materials and ingredients in food processing

2.2 Potatoes

The newest of the great staples to be introduced to the western world, potatoes are boiled, roasted, baked and fried, and are eaten hot or cold (e.g. in salads), and whole, cut in various ways or mashed. Sir Walter Raleigh would have been pleased with what has happened to the vegetable that he reputedly brought back from the New World to Elisabeth I. In contrast to rice, the suitability of different varieties to different functions is not always clear-cut; although a potato for canning or salad use needs to be relatively small and remain firmly intact after processing or cooking, almost any potato can be boiled, and many can be roasted or baked. However, the particular characteristics of each variety and their recommended suitability for each cooking process are now becoming more widely promoted.

Many hundreds of potato varieties have been developed since their introduction to the Western world. As well as the need to produce varieties to suit commercial usage, there has been a significant drive to increase productivity, increase disease resistance and adapt to different growing conditions. Environmental factors are very significant in determining the suitability of particular potato crops for specific end uses. One of the main environmental factors is day length: some varieties produce full crops in response to the long and lengthening days of early summer, others need to grow through to the short days of autumn. In the UK, varieties are classified into

---

Box 5 - Potatoes around the world

There are over 2000 species of potato (genus *Solanum*) in the world, of which eight are cultivated commercially. All but one of these species are limited to regions of South America. The only one to be cultivated throughout the world is *Solanum tuberosum*. The species in South America include several that are resistant to frost, and are grown in the high mountains of Peru and Bolivia. One species, *S. goniocalyx*, is noteworthy for the intense yellow colour of its flesh.

References:

Lisinska, G. and Leszczynski, W. (1989). Potato Science and Technology. Publ: Elsevier Applied Science

Figure 3 - Typical uses of the potato as a raw material and ingredient

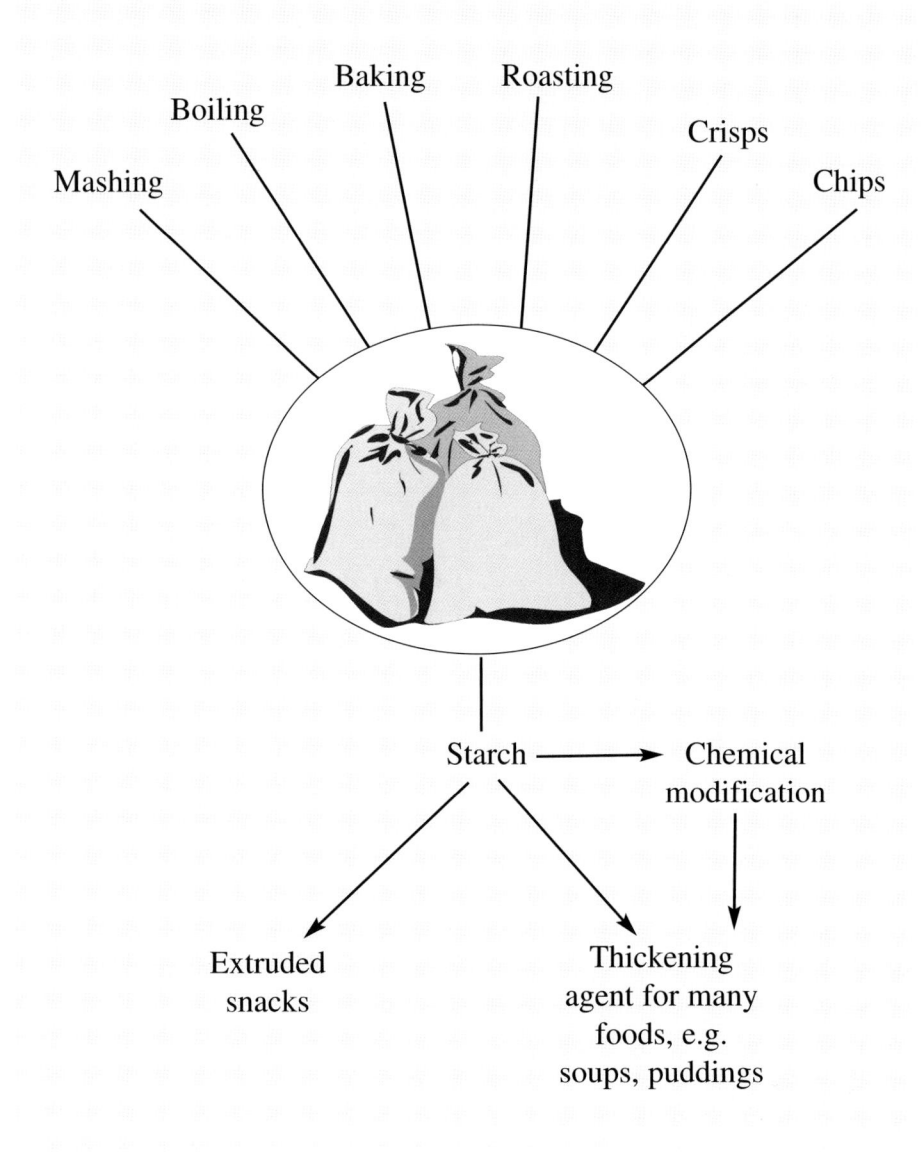

Mashing
Boiling
Baking
Roasting
Crisps
Chips

Starch → Chemical modification

Extruded snacks

Thickening agent for many foods, e.g. soups, puddings

earlies, second-earlies, and early and late main crops, depend on their time of maturity (Cox, 1967). Although agricultural factors have been a major driving force for variety development, it is the suitability for end use that will be focused on here.

The majority of the dry matter content of potatoes is starch, and the content of both of these varies significantly from one cultivar to another and with the growing environment. This variation is generally associated with the length of the growing season, although the related effects like growth temperature are clearly involved, and the type of fertiliser used may also be significant (Thybo *et al*, 2002).

Late potato varieties, which produce a crop into autumn and have a longer growing period, tend to have a higher dry matter content than early varieties. Dry matter and starch contents are, in turn, related to the texture of the cooked product, which itself will depend on the way the potato has been cooked or processed. Factors include the rate of heating or cooling of the potato and the length and conditions of storage of the raw material.

Potato texture changes significantly during cooking; starch 'cells' within the tuber expand and burst, but this disruption is modified by the degree of intercellular adhesion. So the cooked texture results from a combination of the level of disruption caused by the bursting of starch 'cells' and the level of adhesion which remains between starch 'cell' walls. The heating effect also makes the pectic substances in the cell walls more soluble. Generally speaking, a dry matter content higher than 20% is likely to result in a floury open texture in the cooked product and dry matter less than 18% is more likely to give a potato that is waxy and firm. Immature potatoes and early varieties are generally low in dry matter and remain whole after cooking.

Canned and salad ('new') potatoes

The primary requirement in these products is for uniform, small, firm potatoes which retain their shape and firm texture after cooking or processing. In the canning process, it is important to ensure that all parts of the pack are both cooked and sterilised. Small potatoes are desirable as this will reduce the process time required. Uniformity of size will also help to ensure that the processing time is minimised -

a mixture of sizes would mean that extra processing time would be required to cook the larger potatoes - this both increases costs and potentially reduces the nutritional, textural and flavour qualities of the smaller potatoes.

Fried, roasted and baked products

The production of high quality fried or roasted potato products requires a high dry matter content in the raw material. This helps to optimise the texture and fat uptake, characteristics that are desirable in the final product. A high amylose to amylopectin ratio, small cell size and a low sugar content are also significant. Main-crop, floury texture varieties are desirable; frying early season, waxy new potatoes results in a product with a rather oily texture and without the desired fluffy characteristics of a potato chip. As well as texture of the product, final product colour is also important. Critical to the colour of fried products is the reducing sugar content (predominantly fructose and glucose). If too high, this can cause an undesirable darkening in the colour of the chip during frying. The sugar content of potatoes can be high in crops which have either been harvested at an immature stage, or have been in prolonged storage or stored at low temperatures.

Effect of storage on potato quality

Potatoes provide a good example of how storage affects an ingredient's suitability for a specific end purpose. Potatoes that are intended for storage, rather than immediate use, are initially held at 15-20°C and high relative humidity for about 10 days. This is called curing. The two primary effects are that wounds incurred during harvesting are healed and the periderm (the outer skin) becomes thicker. 'New' potatoes ('scrapers') have a very thin periderm. Keeping them in the kitchen for a couple of weeks mimics these curing conditions quite closely and the potatoes will subsequently not be easy to scrape.

Potatoes for table use are subsequently stored at around 4°C and high relative humidity. This reduces sprouting, weight loss and general deterioration. Those intended for baking or frying are stored at higher temperatures (typically 7-10°C)

Raw materials and ingredients in food processing

(see below). At this elevated temperature, potato tubers will begin to sprout during storage. This development is controlled by the application of sprout suppressant treatments, such as maleic hydrazide or CIPC (isopropyl N-(3-chlorophenyl) carbamate).

Chilled storage promotes conversion of starch to sucrose and reducing sugars, which leads to excessive sweetening and non-enzymatic browning on frying (such as unacceptably dark chips, which appear to have been over-fried). If potatoes have been stored chilled, the breakdown of starch can be reversed by subsequent storage for several weeks at 18-21°C (Adams, 2004); this is called reconditioning.

In potatoes which are to be canned or boiled, chilling-induced sugar accumulation may be less of an issue, as the sugars will tend to diffuse out of the potato during cooking. However, if the potato varieties are initially low in dry matter content, excessive loss of starch may result in an unacceptably soggy product.

Varieties are now being developed which show reduced starch breakdown during chilled storage. Modified atmospheres containing less than 3% oxygen can also be used to slow down the loss of starch, but this also tends to increase sprouting (Adams, 2004).

Box 6 - Canning of vegetables

The canning process is, from the vegetable's point of view, relatively exacting. The 'basic' process required is equivalent to the slowest heating point in the can reaching a temperature of 121°C for 3 minutes (called the 'botulinum' cook, as it is designed to cause a one million million-fold reduction in the occurrence of spores of *Clostridium botulinum*, potentially the most significant food poisoning organism associated with this type of product).

This degree of processing is usually more than is required to actually cook the vegetable. Therefore, there is a need to choose varieties that are suited to and can withstand the canning process. As seen with potatoes, it is not just the physicochemical properties of the individual vegetable, but also their uniformity of size. The same is true with vegetables such as carrots. These can come in a

continued....

wide range of sizes, but for canning it is important that they are both reasonably small and uniform in size. However, there are fundamental aspects about the nature of different varieties of vegetables that make one more suitable for canning than another.

Peas are an example of a vegetable whose texture after processing is determined by starch characteristics. The mouthfeel of heat-processed peas depends greatly on starch gelatinisation during processing and also on the extent to which the starch has chemically degraded (Adams, 2004). Peas maintain a higher quality if kept within their pods, and within a particular cultivar, smaller peas tend to be more tender than larger ones (although between varieties there is no correlation - i.e. large seeded varieties are no less tender than small seeded varieties). Having established a suitable variety for the proposed process, it is important that the peas are harvested at the correct stage of maturity. They mature rapidly during growth at high temperatures, and a net conversion of sucrose to starch occurs. It is important to optimise the level of sweetness, starchiness and tenderness. For about 60 years this has been done with an instrument called a tenderometer, which measures the resistance of a pea sample to compression and shearing. Problems with standardising these instruments have led researchers to investigate alternative methods such as near-infrared spectroscopy.

Green or snap beans are a product whose texture is determined by polysaccharides in the cell wall. The dependence on time of harvest of final product quality is particularly important for green beans as the tissues of the bean pod are at different stages of maturity when harvested. The fibrous tissues are at a very early stage of development in the varieties used for processing, whereas the parenchymal tissues are further developed and have relatively thick cell walls (and give the bean its 'snappy' quality). Monitoring maturity at time of harvest is important in preventing excess development of the fibrous tissue and resulting 'stringiness'. High temperatures during growth of the crop can lead to rapid development of fibre and over-maturity (Adams, 2004).

References:

Adams, J.B. (2004) Raw materials quality and the texture of processed vegetables. In: Kilcast, D.E. Texture in Foods. Volume 2. Solid Foods. Publ.: Woodhead Publishing. pp343-363

Box 7 - Cooking and eating apples

The processing of apples into apple juice, canned sauces and slices, and dried and frozen products is a major industry, particularly in the USA. Each product will be affected by the characteristics of the starting material. In the USA, Golden Delicious and Idared among others are processed into apple sauce, whereas Mutsu and Northern Spy are used for slices, and Twenty Ounce is processed into baby food (Downing, 1988). However, in the UK, there is a variety of apple which has been developed for a rather specific use, and which is probably quite unusual compared with apple usage throughout the rest of the world.

The UK is the only country to grow apples specifically for processing and the Bramley apple is the classic cooking apple, accounting for about 95% of all cooking apples sold. Whereas eating apples tend to be sweet and remain fairly firm if processed, Bramleys are significantly larger, much more acid and less sweet, and break down when cooked. This gives the soft, almost mushy product that is desired, especially in home-made apple pies, crumbles and similar products. The first Bramley tree grew from pips planted in a garden in Southwell, Nottinghamshire around 1809 - commercial selling of apples started in 1850. Most Bramleys are now grown in Kent, East Anglia, the West Midlands and in Northern Ireland.

In all foods, flavour is mostly determined by the level of sweetness and sharpness. In apples this is characterised by the balance between sugar and malic acid. Lower levels of acid and higher sugar content in dessert apple varieties give them a sweet flavour, but they lose their 'appley' flavour when they are cooked. Bramley apples, however, are unique because they contain a higher acid content and lower sugar levels to produce a stronger, tangier tasting apple that retains its flavour when cooked. Texture is also important when choosing apples for cooking. Bramley gives a moist, airy 'melt in the mouth' texture whilst dessert apples can produce a chewy, dissatisfying texture.

References:

Downing, D.L.(ed) (1988) Processed Apple Products. Publ.: AVI Publishing

www.bramleyapples.co.uk

Properties of raw materials

> ### Box 8 - Different oranges for different purposes
>
> Oranges are an important crop that are grown in all warmer temperate regions. Sweet dessert oranges (*Citrus sinensis*), the variety that are generally eaten as a table orange, are used for the production of fruit juices and soft drinks as well as being canned as segments. The seville orange, *Citrus aurantium*, is used in the production of marmalade. These oranges are too bitter and too acidic to be eaten fresh, and are principally grown in Seville and Malaga. There is also a limited amount of production in Japan, where it is known as Daidai and is used primarily for flavouring and decorative purposes. The slightly different climatic conditions in the two main growing regions result in different flavour profiles.
>
> Another species of orange, the bergamot (*Citrus bergamia*) is not grown for consumption at all, but for its distinctively perfumed rind oil.
>
> **References:**
>
> Saunt, J. (1990) Citrus Varieties of the World. An Illustrated Guide. Sinclair International Ltd.

2.3 Wheat

Unlike rice and potatoes, the use of the intact wheat grain as a product is somewhat restricted, although several examples do occur. Bulgar wheat is one of the oldest cereal-based foods known and is basically soaked and cooked wheat grains; couscous, which is cooked cracked wheat, is also widely consumed in the Middle East and north Africa, and there are other ancient products, principally from the Mediterranean region. In addition, in the Western World, puffed wheat breakfast cereals have become quite popular, as has whole-grain bread.

However, wheat flour forms the basis of a wide array of food products, and wheat is the most widely cultivated crop in the world. As with potatoes and rice, it is the high carbohydrate content of wheat grain and flour that are most significant in making it important nutritionally, but it is the amount and quality of the protein in the flour that determine the suitability of wheat for end products such as bread, cakes and biscuits. In the UK, protein contents in commercial flours typically range from 9 to 14%. In bread products, the key protein complex of concern is gluten. A vast amount of work has been done on the characteristics of gluten and its components. In this review, the basic concepts will be discussed; in simple terms, wheat varieties high in gluten-forming proteins give strong flour suitable for bread. Weaker flours suffice for cakes and biscuits, where other ingredients have a major effect on final product characteristics, although fruit cakes need flour with bread-type protein characteristics to help prevent the fruit from sinking. The relationship of total protein content, gluten content and gluten quality is complex; the strength of a flour refers to characteristics of the gluten that make it able to hold gas in a dough.

Pasta products are made from durum flour from a separate wheat species (*Triticum durum* (or *T. turgidum* var. *durum*) as opposed to *Triticum aestivum*).

The majority of bakery products in the UK and Western world are based on wheat flour. As will be seen, bread-type products require specific characteristics in certain grades of wheat flour, but even in other flour-based confectionery, the characteristics of the final product are directly related to the nature of the flour.

Bread

Bread is basically made by baking a flour and water dough, often leavened with yeast. The development of the dough correctly to yield a palatable end-product has been a highly skilled craft for thousands of years. Many different types of breads have been developed throughout the world. Some are leavened by fermentation, others are unleavened. Some are baked 'open' (such as French baguette-style products, which need to be eaten on the day of consumption) and others are baked in containers (such as the moister sandwich bread-type loaves that may remain edible for a week or more). It is the interaction of the type of processing given to the dough with the ingredients of the dough that give rise to the wide variety of products that can be produced.

The key factors in the creation of a typical loaf of bread are aeration and leavening: the incorporation of air into the dough, the distribution of this throughout the dough (and ultimately the finished loaf) and the production of carbon dioxide gas during proof or by fermentation to generate the required volume. In traditional British loaves, the aim is to yield a 'risen' product of the desired texture, avoiding the generation of large gas pockets or 'voids' or areas of unrisen dough. (In contrast, products such as ciabatta rely on large gas cells.) The gas therefore has to be trapped within the dough structure and held there when the product is baked (or else the product will collapse); the nature of the dough structure will affect the size and distribution of the gas bubbles. A discussion of the many processes that have been developed for bread making is largely outside the scope of this short book, but it is important to note that the processes developed have significantly influenced the range of flour types that can be used (see Box 9 and 10 on flour blending and the Chorleywood Bread Process).

Although there is more to a fermented bread dough than just flour and water (salt and yeast are also required - and fats, antioxidants and other ingredients may also be used), it is the flour and its protein content that form the basis of the ability to make bread. Wheat flour contains several groups of proteins: albumins; globulins; prolamins and glutelins. It is the prolamins and glutelins that contain the gluten-forming proteins that give wheat flour its almost unique ability to form a dough that can retain gas and increase in volume during the breadmaking process (Cauvain and Young, 2001). In wheat, the prolamins and glutelins are called gliadins and glutenins, respectively, and it is they that are responsible for this property. Bread and other fermented product volumes are directly related to the quantity of protein present in the flour - in general, the higher the protein content, the greater the product volume. Breadmaking flours typically contain about 11.5% protein, whereas 'plain' white flour contains around 9.5% protein.

However, the amount of protein is not the sole determinant of suitability for breadmaking. Individual wheat varieties yield flours with gluten of specific qualities. In particular, gluten quality directly affects the way in which flours will behave when subjected to the stresses and strains of processing (Cauvain and Young, 2001). There are four characteristics that need to be taken into consideration when trying to optimise dough performance:

Raw materials and ingredients in food processing

- resistance to deformation
- elasticity - does the dough return to its original shape after deformation?
- extensibility - how far can the dough be stretched?
- stickiness

The aim is to retain the gas bubble nuclei that have been created during mixing, and to allow them to expand during proving and baking as a result of carbon dioxide produced from fermentation. Gluten is viscoelastic, and it is the balance of its elasticity and extensibility that is important in dough formation. Stickiness tends not to be a significant problem, except when the doughs are moulded or subject to some other form of shearing. In addition, other factors may be important in determining dough and bread quality (Box 11).

Box 9 - Breeding, growing and blending wheat flours for bread use

The suitability of specific flours for making bread is dependent on the genetic make-up of the variety, the conditions under which it is grown and the processes to be used to make the bread. In general, the specific requirements for a particular type of bread or breadmaking process are met by blending different varieties of wheat or flour, often grown in different areas. Each year, the quality of a particular flour will vary, depending on the environmental conditions prevailing in the individual growing area. Thus, identical wheat varieties grown in different parts of the world may yield flours of significantly different characteristics and quality, and this variation is likely to be subtly different each year.

In the UK, the type of blending that needs to be done has changed significantly relatively recently. It used to be the case that white sandwich bread could not be made solely with wheat grown in the UK. For products such as wholemeal and wholegrain loaves, the flour used contained a high level of imported material, particularly from Canada. These often have a protein content of above 13%. For standard bread products the blend used to consist of 65-75% strong imported wheat and 25-35% weaker UK breadmaking wheat ('bakers' flour'). In years with a poor UK harvest, wheats from the EC, particularly France, were used. The resulting flour had a protein content of approximately 12% and a water absorption capacity

continued...

of 62-64%. The level of damaged starch was also an important quality control parameter. When using the Chorleywood Bread Process (see Box 10), a larger proportion of weaker gluten wheats could be used, with a slightly lower overall protein content (11-11.5%). The advantage to the baker was that such blends were cheaper and it reduced the reliance on imported wheats, although there was always controversy as to whether this affected the final product quality.

Since the 1970s, the need to go abroad for a supply of wheats strong enough to make bread has diminished. This change was driven by the EC Common Agricultural Policy, which put a significant premium on buying wheat from outside of the Community. There is currently a limit of 5% before the premium is levied. The result was that much more research and development was put into developing the varieties that could be grown in the UK to improve their suitability for producing breadmaking flours. Now that the flours from these varieties have improved characteristics, the need to blend UK flour with that from abroad has largely disappeared. However, blending *per se* is still required, and a small level of largely Canadian wheat is still used in the industry. Different blends will still give characteristics suited to particular types of bread, and there is still the factor of variations in growing conditions from year to year.

References:

Brown, J. (1985) (ed.) The Master Bakers' Book of Breadmaking. 2nd Edition. Publ.: The National Association of Master Bakers, Confectioners and Caterers.

Box 10 - The Chorleywood Bread Process

The design of a food production process can have a significant effect on the number and level of ingredients that go into the formulation: if you change the process, you may have to alter the formulation to get the same end product. The fundamental key to bread production is the development of the gluten structure in the dough. Before the early 1960s, the most important stage in this was an initial bulk fermentation stage in the dough, which complemented the development achieved by the actual mixing of the dough, as well as producing small levels of fermentation products that contributed to the specific flavour of

continued...

bread. It was commonly known that a short period of intensive mixing of the dough, which effectively means the input of extra mechanical energy into the dough, could mimic some of the effects of the bulk fermentation stage, and various experiments were carried out with continual mixing of the dough to try and optimise gluten development in the dough. Too much mixing or too long a bulk fermentation stage resulted in a fall-off in quality of the dough and subsequently the bread produced.

The key factor in the Chorleywood Bread Process was that it was found that optimal dough development could be achieved by the control of the initial mixing work input - to 39.6kJ/kg. It was also found that this needed to be achieved in no more than 2-5 minutes. Longer times resulted in opening of the crumb grain structure in the final loaf. This necessitated the use of high-powered mixers that were only just becoming generally available in the UK at the time.

As well as other modifications to the process (the actual mixing action, and the application of positive or negative pressures had an effect on dough development), the change in process required modifications to the formulation. It was found that rapid chemical oxidation was required in dough. This meant the addition of either a low level of a fast-acting oxidant such as potassium iodate, or higher levels of a slower-acting agent such as potassium bromate or ascorbic acid. Only ascorbic acid is currently permitted in the UK and EU and a level of 75ppm of flour weight was recommended. Also required was the addition of fat to the dough, at about 0.7% of flour weight, as well as extra water to soften the dough and more yeast (a 50-100% increase depending upon the scale of the production facility).

Bread from the Chorleywood Bread Process was found to be softer and whiter than that produced by bulk dough fermentation, and although it was feared that the lack of a pre-ferment stage might result in a loss in flavour of the final loaf, taste panel tests could not detect any such difference.

The whole breadmaking process could now be controlled more accurately and objectively, with considerable savings in time and space.

References:

Chamberlain, N., Collins, T.H. and Elton, G.A.H. (1961) The Chorleywood Bread Process. BBIRA Research Report No. 59

Chamberlain, N., Collins, T.H. and Elton, G.A.H. (1962) The Chorleywood Bread Process. Bakers Digest **36** (Oct): 52-53

Properties of raw materials

Box 11 - Non-protein factors that affect flour quality

The level of α-amylase in the flour is a factor that can affect the quality of the final product. The enzyme catalyses the hydrolysis of starch into lower-molecular-weight dextrins. If the formation of dextrins during the breadmaking process is too great, this can lead to problems in slicing the bread. In bulk fermentations (see Box 10), high cereal α-amylase levels can also lead to a softening of the dough, which will affect the processing of the dough and final product quality.

α-Amylase can also be a factor if there is excessive starch damage in the flour, as it is more susceptible to attack by the enzyme. During wheat grain milling, some of the starch granules can be subject to high pressures, and their surfaces may become mechanically ruptured or damaged. One of the key formulation calculations in breadmaking is the ratio of water to flour in the recipe. Damaged starch absorbs approximately twice its own weight of water, whereas undamaged starch only absorbs about 40% of its weight. As the degree of water absorption is a critical element in dough development, it is important to control the level of damaged starch in the flour. Excess levels of starch damage can lead to a greying of the crumb colour, and a more open cell structure - i.e. the bubbles in the bread have a larger average size.

References:

Cauvain, S.P. and Young, L.S. (2001). Baking Problems Solved. Publ: Woodhead Publishing

Cakes

Cake formulations are basically batters based on flour, eggs, sugar and fat (butter or margarine). Unlike with bread, there is no formation of an aerated dough and protein content of the flour is not so critical. The aeration of the final product comes from the fine bubble structure in the batter and the production of carbon dioxide from sodium bicarbonate (baking soda) incorporated into the mix (either as a separate ingredient - see Section 3.11, or by the use of 'self-raising' flour). The basic recipe for a cake is the 'pound' formula in which equal quantities of the four main ingredients are used, and virtually any type of flour will suffice. This recipe is

Raw materials and ingredients in food processing

known as 'low ratio' and the resulting cakes have a limited shelf life, a coarse texture and a tendency to dry quickly. Increasing the sugar content increases shelf-life and moistness, but without modification to the flour, the phenomenon of cake collapse occurs on leaving the oven, due to insufficient setting of the structure. This potentially limits the use of extra sugar to improve 'eating quality'.

In 1931, a mill chemist called Montzheimer noticed that treating flour with chlorine resulted in cakes with a finer texture and bigger volume. The properties of the chlorinated flour also allowed the development of 'high ratio' cakes, with sugar levels as high as 140% of flour weight and equally high liquid levels. The resulting product had a longer shelf life, a finer texture and a more moist crumb. This type of product became widespread in Britain and Ireland, Australia, Canada, the USA and Japan, but flour chlorination was never permitted in much of mainland Europe.

Table 1 - Typical high- and low-ratio cake formulations

Ingredient	High-ratio cakes (% of flour weight)	Low-ratio cakes (% of flour weight)
Flour	100	100
Caster sugar	115	100
Shortening (fat)	60	100
Milk powder	7	0
Salt	2.5	0
Baking powder	4	0
Water	70	0
Whole egg	80	100
Glycerol	8	0

The beneficial effects of the chlorination of flour seem to derive from effects on the starch granule fraction, with the granule surface becoming more hydrophobic. This, together with a stronger interaction with egg protein in the batter, results in an increase in the viscosity of the batter during the early stages of the baking process, which helps to maintain the fine bubble structure of the batter. As the batter sets, the steam pressure within the bubbles continues to increase, eventually causing them to burst and create a continuous structure, in effect changing it from a foam to a sponge. As the product cools, the pressure within the bubbles decreases and air is allowed to filter in to balance the internal and external pressures and so avoid collapse. With an untreated cake flour, although the cake volume is at least as great in the oven, the structure sets later. Thus, it remains as a foam for longer and tends to collapse on cooling.

Since the introduction of 'high ratio' formulations, much development has gone into the improvement of product quality by modifying the individual recipes, and by trying to analyse at a microscopic level what physical characteristics in the flour were important. The incorporation of shortenings with added emulsifiers was very successful, and flour particles of 15-25µm seemed to be optimal, with a controlled amount of starch damage. The result was a wide range of very popular products based on the use of chlorinated flour, with each ingredient having an important role to play and with the relative amounts of each ingredient and their interactions being very important.

However, pressure began to grow on the industry to demonstrate that the use of chlorine in flour treatment was actually safe (i.e. that there would be no adverse health effects from eating the treated flour in cakes). Although the industry was confident that there were no significant problems, it was felt that consumer pressures and legislative requirements would eventually lead to a decline in the use and acceptability of the treatment, and so it decided not to pursue food safety trials. As a result, by default, the use of chlorinated flour has not been permitted in the UK since November 2000. Leading up to this, the industry was faced with the option of either reverting to low ratio formulae or developing an alternative method for producing high ratio cakes. The former option was quickly rejected as it was felt that the significant change in and lowering of product quality would not be acceptable to consumers. In order to be able to continue to produce high ratio cakes, it was therefore necessary to either find an alternative method of treating the flour, modify recipes, or do some combination of the two.

Although acetylation of flour using ketene or acetic anhydride has been considered, heat treatment has been the area in which most work has been carried out in the treatment of the flour. Among the ingredients that have been considered are various types of starch (wheat, oat, potato, maize, rice), oat fibre, xanthan and guar gums, emulsifiers such as lecithin and proteins such as whey, as well as variations on existing ingredients such as the egg (Catterall, 2000). Although the decision to use heat-treated flours to produce high-ratio cakes was fairly straightforward, as the products obtained were of similar quality to those obtained by using chlorinated flour, the complex set of interactions that occurs between ingredients means that there will always be subtle differences. There may also be examples where the new product made with heat treated flour just cannot match that of the 'old' product that used chlorinated flour. This has meant that much research and development has had to be put into optimising product quality by subtly changing ingredient types and levels. In work at CCFRA (Cook, 2002), it was found that cake specific volume, a major measure of product quality, was greatly influenced by the water level in the recipe, although the characteristics of the flour and the baking powder type were influential as well. Baking temperature and flour/sugar combinations were also significant. Statistical models showed that maximum specific volume would be obtained with:

- a low level of water
- flour with lower protein/higher starch levels and a favourable particle size distribution
- baking powder with a slow rate of reaction

Heat-treated cake flour was also found to gelatinise at a lower temperature than found with chlorinated flour, and less completely.

Pasta

Why does pasta generally require the use of durum wheat flour rather than cake-making flour? The reasons are numerous, complex and often subtle. In simplistic terms, durum wheat has more of the characteristics suitable for manufacturing pasta products than does bread flour. Pasta and bread are very different products made from very similar starting materials - predominantly wheat and water. It is unsurprising that the optimum characteristics of the flour starting material are different for the two.

As with bread wheats (*Triticum aestivum*), durum wheats (*T. durum*) vary widely in their characteristics, depending on the variety and the environmental growing conditions. It was originally noted that the wheats grown in southern Italy and north Africa made greatly superior pasta than the softer European wheats. The former were durum wheats. Most durum wheats are amber in colour (although red varieties, which are mainly used for animal feed, also exist). This helps to give pasta products their familiar final colour; this was originally associated with a high-quality product - pasta had to be yellow to be good, but this association no longer holds. Durum wheats are also the hardest wheats known, which gives them specific milling characteristics.

Durum wheats are usually considered to have higher protein contents than bread wheats, but this may be mainly due to the environmental conditions under which they tend to be grown. Most durum wheat growing areas undergo severe weather conditions, and so the potential grain yield and quality depends largely on the ability of the variety to adapt to different conditions (Troccoli *et al*, 2000). The semi-arid conditions which often prevail tend to result in low yield of high-protein wheat. This suggests that the development of pasta products may have been linked to the quality of wheat which could actually be grown successfully in individual areas of Europe and the Far East (the two main centres of pasta product development). Interestingly, whole grain type products such as bulgar and couscous, which are also best made with durum wheat also originate in the Mediterranean and north Africa, where such varieties are best suited to the environmental conditions.

Although protein levels tend to be higher than those of bread wheat, gluten quality varies quite widely, and durum wheats tend to be less strong than bread wheats grown under similar conditions.

The types of differences found between bread and durum wheats can be compared with those found between different types of rice. Certain characteristics of durum wheat interact to give a better pasta product than can normally be obtained from non-durum wheat. Durum wheats tend to have larger kernels, which are longer in relation to their height and width than common wheat. The endosperm tends to be higher in ash (mineral) content. Although the whole wheat kernel does not show this enhanced ash content, because of the hardness of the kernel and therefore the yield of flour and semolina from the milling process, these flours and semolinas can contain 50% more ash than comparable common wheat flours.

Pasta manufacture is essentially a very simple process: water is added to flour, meal or semolina and mixed into a very stiff dough. This is then shaped into the desired form and either cooked immediately and eaten or dried for later consumption. The best pasta is made from semolina, rather than flour. Semolina is basically chunks of endosperm that have not been ground down into a flour and which maintain their structure when mixed with water. Semolina with smaller particle dimensions are preferred for pasta, whereas larger semolina is ideal for milky puddings. Durum wheats, because of their endosperm characteristics, give a greater yield of semolina than do other wheats, and the resultant stiff doughs are best suited to modern pasta making processes. These doughs are stiff and stable, but flow readily under pressure (as in extrusion into spaghetti strands) and do not show the elasticity found in stronger bread wheats. The best milling quality is found with large, vitreous kernels of uniform size, with the wheat being high in protein content and with medium-strong gluten characteristics. With pasta made from flour rather than semolina, these milling characteristics become less important. More details on the specific properties in the wheats that millers and processors look for can be found in Irvine (1978) and Dick and Matsuo (1988).

The most important difference from the product point of view is that durum wheat yields a product that is much more stable when cooked; it does not disintegrate to the same extent as common wheat pasta would, and tends not to become as mushy if kept in water after cooking, or if canned (Irvine, 1978).

Box 12 - Reformulating for coeliacs

Coeliac diease, also called gluten-sensitive enteropathy or gluten intolerance, is probably the most widespread food intolerance condition in the UK. It is caused by dietary wheat gliadins and related proteins found in other cereals, including spelt wheat - the wild ancestor of modern cultivars of wheat. Gliadins are one of the major fractions of the wheat storage protein, gluten (glutenins are the other), which is the central factor in the production of wheat dough, and subsequently bread and similar products. Coeliac disease is a cell-mediated allergic reaction. This differs from typical hypersensitivity allergies (e.g. nut allergies) in that it does not involve the rapid release of immunoglobulin E. Symptoms usually take significantly longer to develop, typically 6-24 hours, and peak after about 48 hours before beginning to subside (Taylor and Hefle, 2001). Although minor traces of gliadins in foods (e.g. through cross-contamination) are not the major issue that they are in immunoglobulin E-mediated allergies, people with coeliac disease do have to avoid products containing wheat, barley, rye and possibly oats (there is conflicting evidence of the role of oats in coeliac disease and this may vary from one individual to another).

The industry has a three-pronged problem:

- to try and formulate some bakery products that are of acceptable quality yet suitable for coeliac sufferers,
- to ensure that both 'suitable' and 'unsuitable' products are appropriately labelled, and
- to ensure that 'suitable' products are not contaminated with 'unsuitable' ingredients.

Product reformulation is a major technological problem. Wheat is a major ingredient in a variety of products, from breakfast cereals, through biscuits and cakes, to bread and pastries, and pasta. There are literally thousands of products which are not suitable for coeliac sufferers, and which would be candidates for reformulation with gluten-free flour and related ingredients. Not only does a product reformulation have to produce something that is edible and acceptable, it also has to have a reasonable shelf-life, and be economical to produce. The production runs are likely to be smaller, and unit costs will probably be higher, even after the development work has been completed. Despite this, the number of gluten-free alternatives that have been investigated are numerous. As has been demonstrated above, the characteristics of wheat proteins are fundamental

continued....

in the production of many cereal-based foods, particularly bread and cakes, and producing similar products without wheat is a complex task.

Some products recently available through major retail outlets include:

- cake mixes from Nutrition Point Ltd, which have rice flour and potato starch as principal ingredients;
- rice cakes from Kallo;
- an organic bread from The Village Bakery, based on maize flour and chestnut flour;
- a lemon cake from UltraPharm Ltd, made with gluten-free wheat starch;
- a selection of biscuits from Nutricia Dietary Products based on maize starch and flour, soya flour and tapioca starch

Dean (2000) lists some wheat- and gluten-free products available on prescription and over the counter; these include pasta products, bread products and pizza bases, flour mixes and biscuits and crackers. The University of Saskatchewan (1999) recently patented an extruded pasta product based on pea flour, while Caperuto *et al,* (2001) looked at the use of quinoa and maize flour mixtures. Rice varieties have been examined for their suitability for making bread (Torres *et al*, 1999), and cassava flour has been evaluated for its use for muffin production (Chauhan *et al*, 2001).

References:

Caperuto L.C., Amaya-Farfan J., Camargo C.R.O. (2001) Performance of quinoa (*Chenopodium quinoa* Willd) flour in the manufacture of gluten-free spaghetti. Journal of the Science of Food and Agriculture, **8** (1): 95-101.

Chauhan, S., Lindsay, D., Rey, M.E.C., and von Holy, A. (2001) Microbial ecology of muffins baked from cassava and other nonwheat flours. Microbios, **105** (410): 15-27.

Dean, T. (2000) Food intolerance and the food industry. Publ: Woodhead Publishing.

Taylor, S.L. and Hefle, S.L. (2001) Food allergies and other food sensitivities. Food Technology, **55** (9): 68-83

Torres, R.L., Gonzalez, R.J., Sanchez, H.D., Osella, C.A., and de la Torre, M.A.G. (1999) Performance of rice varieties in making bread without gluten. Archivos Latinoamericanos de Nutricion, **49** (2): 162-165.

University of Saskatchewan (1999) Production of legume pasta products by a high temperature extrusion process. United States Patent 5,989,620.

Properties of raw materials

Table 2 - Reformulation effect on nutritional content

Reformulating a product so that it suitable for coeliacs does not necessarily significantly alter the nutritional quality of the food, as the figures below illustrate.

Nutrient	*Standard product*	*Gluten-free product*
Energy (kJ/100g)	2151	2115
Protein (g/100g)	5	3.5
Fat (g/100g)	25.9	25
Carbohydrate (g/100g)	69.2	66.5

References:

Holland, B., Unwin, I.D and Buss, D.H. (1988). Cereal and Cereal Products. Third supplement to McCance & Widdowson's The Composition of Foods. 4th Edition. Publ: Royal Society of Chemistry

NewFoods (1999) A CD-RoM database of new products purchased in the UK 1997

2.4 Conclusions

The characteristics of the major cereal, vegetable and fruit products of the world have been tailored by breeding to better meet the desired end-product characteristics. This is particularly true for three of the world's major staples: rice, wheat and potatoes. However, this has not simply been a case of developing a series of variations from one starting point. It is a complex web of variations in environmental conditions (such as temperature and amount of rainfall), agronomic practices (time of harvest, length of storage etc) and preparation and processing conditions that have interacted with the development of different varieties for specific purposes. In simple terms, farmers learned what would grow in their own back yard, and how varying harvesting and processing regimes could produce different end-products. As the food industry became increasingly more global, different ideas from different regions could be exchanged and novel products produced as a result. Although this section has dealt solely with plant-based products, dairy products have also, to a lesser extent, gone through this development cycle, and Box 14 serves as a useful example of this.

Box 13 - Mimicking gluten functionality in bread products

Because bread structure formation relies so heavily on the unique properties of wheat gluten, producing gluten-free bread has always been one of the bigger challenges facing bakery technologists and researchers. A combination of ingredients is required which can mimic the gas-retaining properties of a gluten dough. Although such products have existed for some time, they have generally been thought of as being poor in quality and having noticeably different eating characteristics to conventional bread. Recently, research at Teagasc (the National Food Centre in Ireland - www.teagasc.ie) has developed a new formulation which, it is hoped, will provide a high-quality bread product for those with gluten intolerance. A combination of potato starch and rice flour had been used to improve the taste, texture and volume of the bread. Two hydrocolloids, xanthan gum and HPMC, a derivative of cellulose, were also used to help bind the other ingredients together, mainly because of their strong water binding abilities.

The new formula, which has taken the Teagasc team three years to perfect after experimenting with a number of different starches and proteins, could provide an opportunity for bakers to take advantage of a growing niche market. Recent figures in Ireland estimated that one in every 150 people may have coeliac disease, and a recent medical study on the UK claimed that one in every 100 to 200 people may have the disease but be undiagnosed. Many more may have milder intolerances causing uncomfortable symptoms such as headaches and abdominal pains.

References:

http://foodnavigator.com/news/

Box 14 - Milk variation effects on cheese characteristics

Although milk from different sources (i.e. cows, sheep and goats) is equally suitable as a drink from a functionality point of view, some of the inter-specific differences do have a significant effect on the characteristics of cheeses made from them, and also affect the type of processing conditions that have to be employed. Within cow's milk, too, there are breed-related differences and variations due to seasonal, lactational and nutritional factors (Fox, 1993a). Among the major differences of importance are the concentration and types of casein found, the concentration of fat and especially the fatty acid profile, and the concentration of minerals and salts, especially calcium. The interaction between these factors can be complex. Syneresis is an important stage during curd formation - this is where water and dissolved components (whey) are expelled because the curd contracts. The degree and rate of syneresis can affect end product quality and characteristics. Higher fat content tends to slow syneresis, and variations in calcium content in the milk has a marked effect. Goats' milk shows greater syneresis than cows' milk, and ewes' milk shows the least.

Ewes' milk has more than twice as much fat and protein as cows' milk. One of the consequences of this is that cheese yield is about double that from cows' and goats' milk. However, the high antibacterial activity is a problem if fresh milk is used for cultures. Differences in total casein content and the ratio of alpha and beta casein types also result in cheeses of different characteristics, the two best known being Roquefort and Feta. The higher casein content means that a relatively smaller quantity of rennet is used for the same coagulation time.

References:

Fox, P.F. (1993a) Cheese: Chemistry, Physics and Microbiology. Volume 1. General Aspects. Publ.: Chapman & Hall

Fox, P.F. (1993b) Cheese: Chemistry, Physics and Microbiology. Volume 2. Major Cheese Groups. Publ.: Chapman & Hall

3. SPECIFIC INGREDIENTS

A very wide range of ingredients are used in formulated foods. All have their own role to play. It would be impossible to mention more than a small percentage of these ingredients, so this chapter will look at the major groups of ingredients that have a functional role and explain the main reasons for their presence. Within each group, there is much variation and choice, with very different ingredients performing similar roles, but with often subtly different properties. It is important to ensure that the choice made is suitable for the individual application. This requires the user to understand what is responsible for each characteristic and how it contributes to final product quality. All of these ingredients interact with each other - increasing one may require the presence or absence of another ingredient, or a variation in relative levels. As a simple example, adding fruit to a cake mix may mean that ingredient levels in the mix have to be altered slightly for optimum product quality, and that a stronger flour has to be used to stop the fruit sinking to the bottom of the cake.

One manufacturer's ingredient is another manufacturer's final product. Much effort and research goes into identifying new ingredients and elucidating their interactions during processing and storage. Hopefully the examples given will help to indicate the complexity of developing food product formulations.

3.1 Fats and fat substitutes

Fats and oils, collectively known as lipids, are not only essential components of human and animal nutrition, they also provide the food manufacturer with an excellent group of materials to include in formulated foods to give a variety of texture and mouthfeel characteristics. The main form in which lipids are used is as triglycerides (see Box 15).

Specific ingredients

Box 15 - Diagrams of triglyceride and fatty acid structures

The carbon atom has a valency of 4, which means that it has 4 'arms' with which to bind to other atoms. Triglycerides are made from glycerol with three fatty acids attached. In a triglyceride in which all of the available 'arms' of the carbon atoms in the fatty acid portion are linked either to neighbouring carbon atoms or to hydrogen atoms (i.e. a saturated fatty acid), the following structure is obtained, where 'x', 'y' and 'z' may indicate the same or different numbers of carbon atoms (usually 14, 16, 18 or 20):

$$\begin{array}{l} H_2C - OH \\ | \\ HO-CH \\ | \\ H_2C-OH \end{array}$$

Glycerol

$$\text{(Fatty acid) } CH_3 - (CH_2)_y - \overset{O}{\underset{\|}{C}} - \underbrace{\begin{array}{l} CH_2 - O - \overset{O}{\underset{\|}{C}} - (CH_2)_x - CH_3 \text{ (Fatty acid)} \\ | \\ O - CH \\ | \\ CH_2 - O - \overset{O}{\underset{\|}{C}} - (CH_2)_z - CH_3 \text{ (Fatty acid)} \end{array}}_{\text{derived from glycerol}}$$

Fatty acids themselves are usually depicted as angled chains, with each angle depicting a carbon atom (and its associated hydrogen atoms) - palmitic acid is shown. In esterification to form a triglyceride, the OH's from each of the fatty acids and the H's from glycerol OH's are lost as 3 water molecules.

Carbon has the ability to make double bonds with itself, with two hydrogen atoms being excluded. Fatty acid chains with one double bond are termed monounsaturated, while those with two or more double bonds are polyunsaturated.

The tryglycerides with shorter-chain, unsaturated fatty acids exist as light oils, whereas those with longer-chain saturated fatty acids are solid fats and lards. By mixing different ingredients, a range of products can be derived with very specific textural and functional properties. Chemical hydrogenation can also be employed to produce saturated fatty acids from unsaturated starting materials. This is useful, for example, if trying to produce a vegetarian product that requires the functionality of saturated fats in its formulation, as these are more associated with animal products. However, hydrogenation is rarely taken to the extreme of all fatty acids being converted from unsaturated to totally saturated. Some mono-unsaturated fatty acids will remain, and some of these will be in the 'trans' form, rather than the 'cis' form, i.e. the hydrogen atoms will be on the opposite side of the double bond, rather than the same side. Trans fatty acids have been linked with heart disease in recent years, and although the exact relationship is not clear-cut, the food industry has responded to consumer concerns by trying to find alternative ways of achieving the correct blend of fats in product to achieve the desired result.

Different products require fats or oils with different characteristics; the level of fat used in a product can also have a marked effect on product characteristics. The major role of fats in bread is in facilitating gas retention in the dough. The amount that is needed varies with flour type, higher levels being required with stronger white flours, and even more being required for brown and wholemeal flours. The fat probably acts by controlling gas bubble size and stability. It is known that it is only the solid portion of the fat that is functional, so it is important to use a fat that stays solid throughout the proving phase. In the Chorleywood Bread Process, this means having a fat that has a melting point above 45°C (Cauvain and Young, 2001). The level of fat used is also significant in the texture of the final product - higher levels are used in products such as soft bread rolls. A good example of how changing from one type of fat to another can change the nature of the product is to look at the use of oils rather than fats in speciality breads such as ciabatta. Here the oil is added after the proof stage - the dough rises and then collapses during proving; olive oil is the preferred oil for its flavour - it is not added for maintenance of dough structure.

The nature of the fats or oils can also significantly affect the final quality of products such as mayonnaise (see Section 4.5).

Box 16 - Fat substitutes and mimetics

The ways in which fat levels can be reduced will vary from product to product. In some areas it has so far proved impossible to produce reduced or low fat products that are acceptable from a quality point of view and commercially viable. However, there are many products in which fat levels have been reduced by the use of substitutes or mimetics (i.e. components which mimic the behaviour of fats).

Traditional ways of reducing fat levels in formulations have included using emulsifiers, substitution by air or even water, using reduced-fat ingredients (e.g. skimmed milk in place of full-fat milk or cream), and baking products instead of frying them. However, most of these operations have limited applications, and much work has gone into developing fat substitutes and mimetics.

Several types of sucrose polyesters have been developed. These consist of a central sucrose molecule esterified to 6-8 fatty acid units.

By modifying the fatty acid chain length and degree of saturation, they can potentially be tailor-made to simulate fat in baked and fried goods. They are generally stable at cooking, frying and baking temperatures, which makes them technically useful for bakery products. As the 'outside' of the molecule is fat-based, they function as fats in products, but their large molecular size means that can not be digested and absorbed into the body (they are too big for the digestive enzymes to act upon) and so contribute no calories to the diet. Examples of these include Olestra (from Procter & Gamble) and Salatrim (developed by Nabisco). Sorbestrin, from Cultor Food Science, is based on the polyol sorbitol, again esterified to several fatty acids. Unfortunately, from a technology point of view, these types of products have generally not gained widespread regulatory approval, in some cases because of lingering health concerns.

Emulsifiers can be used in relatively small amounts to replace fats, or to allow less fat in a product to perform the same technological function. Work at the Flour Milling and Baking Research Association (Cauvain *et al*, 1988) showed that glycerol monostearate could allow the fat level in a high-ratio cake formulation to be reduced from 18 to 9% of the cake weight without any change in eating quality being detected. However, these emulsifiers are themselves fats and so care has to be taken in making 'reduced-fat' claims.

continued....

There are many fat mimetics available on the marketplace, including over 40 based on starch. These form a gel which holds water and mimics the textural characteristics of fat. They are used to increase viscosity, bind and control water, and contribute to a smooth mouthfeel in fat-replacing systems. They have to be able to withstand food processing operations and still provide the required texture in the final product. Native starches from corn or wheat can be used as fat replacers in biscuits and cakes, whereas pre-gelatinised, modified, high-amylose starch is suitable in baked goods, icings and fillings. As they are used to achieve functional and sensory properties, they perform best in higher moisture foods such as cakes, but generally less well in biscuits and crackers.

Other fat mimetics that have been developed and utilised include maltodextrins, protein-based products and fibre/cellulose-rich ingredients. The latter bind water within their structure, helping to give the impression of fat. Oatrim is one example; it is made by partial hydrolysis of an oat or cornflour bran fraction and can be added as a dry powder or a hydrated gel. It gives a buttery mouthfeel to the final product. Protein-based mimetics include Simplesse (from NutraSweet Kelco), a microparticulated protein. These are formed by hydrating egg white and/or milk proteins and subjecting the mixture to a heating and blending process. Further processing blends and shears the gel to form microscopic, coagulated, deformable particles that mimic the mouthfeel and texture of fat. Although many protein-based fat mimetics are not heat-stable enough to withstand frying, they are suitable for dairy products, salad dressings and products that may be cooked.

There is no single fat replacer that fits all applications. Fat is a complex, highly functional ingredient, and can only be replaced by a combination of other ingredients.

References:

Catterrall, P.F. (2001) Fat replacers and substitutes. Challenges in the development of reduced fat bakery products. In: 'New Technologies - The Future Today'. CCFRA Symposium Proceedings

Cauvain, S.P., Hodge, D.G. and Screen, A.E. (1988). Changes in the fat component of cakes and biscuits to meet dietary goals. Part 1. Fat reduction in cakes. FMBRA Research Report No. 140

Specific ingredients

> **Box 17 - Reduced fat bakery products**
>
> Work at CCFRA in the late 1990s (McEwan and Sharp, 1999; Sharp, 1999) demonstrated the problems associated with trying to reformulate bakery products with a reduced fat content. Although there are a number of replacers and alternative systems available, for many products it has proved both technically difficult and expensive to produce acceptable products. A survey of manufacturers indicated that there were insufficient ingredients available to produce the new products and that eating quality and shelf-life of what was currently being produced was poorer than for 'standard-fat' recipe products. A lack of knowledge of the technical functionality of alternative ingredients was a major problem. Whilst small reductions in fat content were feasible in some cases, the need for a 25% reduction before a 'reduced fat' claim could be made was often too much, and thus one economic justification was lost. Alternative ingredients were generally more expensive and manufacturers found that not enough customers were willing to pay the extra to make new product development worthwhile.
>
> A survey of consumers suggested that they perceived bakery items as treats and indulgences, often for special occasions, and that small reductions in fat content were not seen as any great benefit. There was also the perception, often from past experience, that the lower-fat items were of poorer eating quality, and that they were more expensive. One solution suggested for this was for completely new lower-fat products to be developed that formed a category of their own, rather than being newer versions of existing products.
>
> **References:**
>
> McEwan, J.A. and Sharp, T.M. (1999) Barriers to the consumption of reduced fat bakery products: final report. CCFRA R&D Report 85
>
> Sharp, T.M. (1999) The technical and economic barriers to the production of reduced fat bakery products. CCFRA R&D Report 86

3.2 Emulsifiers and stabilisers

Most lipid (fatty) ingredients are not water-soluble. This poses a fundamental problem in food formulation: how to mix to the fatty components with the non-fatty components in a liquid or dough, e.g. in a cake mixture. What is required is something that will mix with both and therefore allow the two groups of components to be brought together, in exactly the same way that soaps and detergents are used to dissolve grease in water, a process that is called emulsification. The most widely used food raw material for emulsifying is egg yolk, which contains high levels of lecithin (also called phosphatidyl choline), a type of phospholipid. In simple terms, phospholipids comprise a fatty acid chain at one end that is hydrophobic (or lipophilic - i.e. it is fat-soluble and not water-soluble), and a hydrophilic (water-soluble) phosphate-based group at the other end. The fatty acid chain can vary in length and in the degree of saturation, so lecithins extracted from different sources may have slightly different characteristics.

Eggs are widely used in bakery and related products for their emulsifying properties, but clearly there will be many products in which the use of eggs *per se* is inappropriate (because of their flavour, colour or other properties). Lecithins extracted from eggs are, therefore, used extensively in a wide range of food products - from ice-cream and other desserts to corn snacks.

Lecithins can also be extracted from other sources. The second major source is soya. There are some groups of consumers for whom it is important to know the exact source of the lecithin. There are a growing number of people who suffer from food allergies to a greater or lesser extent. Both eggs and soya are major food allergens, and as such have been included in new European Union-derived food labelling regulations that require the presence of potential allergens to be given prominence on food labels. In both cases, and with all known food allergens, it is the protein content of the egg or soya that triggers the allergic reaction. Therefore, purified lecithins from either source should not pose a problem. However, as even minute traces of protein in the product could result in allergic response in an individual, those with severe allergies will need to know that egg or soya derivatives have been incorporated into the product, so that they can decide whether or not to purchase and eat the food. The European Food Safety Authority has been evaluating various derivatives of allergenic ingredients to determine which can be exempted from the specific 'allergen warning' labelling, because the risk is negligibly small. Egg and soya lecithin are currently not exempt.

Another issue which the food manufacturer needs to consider is whether the product is designed to be suitable for vegans to consume. Vegans do not eat anything that is derived from animals; clearly, egg lecithins are animal-derived and so if the rest of the product is suitable for vegans, the manufacturer may consider using soya lecithin as an alternative. However, if there are other animal-derived ingredients in the product, this ceases to be an issue.

To add a further complication on whether to use egg or soya-derived lecithins, the issue of genetic modification has become a major factor. Soya is one of the crops that have been the focus of much research and development in this area - GM soya is widely traded, and some consumers keep a close watch on the use of soya-based ingredients in food products. They may be more tempted by the product if it uses egg-derived lecithin (unless, of course, the eggs come from hens that have been fed GM corn! - this is not currently permitted in the UK).

3.3 Starch and related thickeners

Starch in its many forms is probably the most widely used functional ingredient in food formulations. Starch is the major component of flour. Because starch and flour are major nutritive ingredients in a range of products, it is sometimes difficult to separate their role as a food staple from their role as a functional ingredient in product formulation.

The main characteristic of starch is that it swells and changes structure when heated in water. Raw starch is basically indigestible and the need to cook these staple foods would have been known to early mankind. This would have probably led quite quickly to the knowledge that, when cooked, flour had significant thickening properties.

Starch is the form in which plants store their carbohydrate reserves. As such, the starch from many different plants, or indeed the plants themselves, can be used for its thickening properties in a variety of foods. Starch occurs in water-insoluble granules in plants and is a polymer of glucose that occurs in two forms: amylose and amylopectin. As has been described in Section 2.1 on rice, small changes in relative levels of these two forms can contribute significantly to the functional characteristics of the rice, and consequently the uses to which it can be put. Starches from different plants show even greater variation and this may be used to produce a range of functional characteristics. However, all starches have the same basic characteristics. When the insoluble starch granules are suspended in water and heated, they absorb the water, swell and burst, releasing the starch and forming a viscous solution that gels on cooling. Over time, the straight chain, medium length amylose molecules in the gel tend to become oriented together. This results in a type of crystallisation effect and the starch precipitates out of solution. As a result, the gel weakens and eventually breaks, and water is lost from the gel (called syneresis). This whole process is known as retrogradation.

Gelling and retrogradation characteristics vary significantly among starches. Starches in which the amylopectin molecules have longer chain lengths (e.g. potato as opposed to maize) are less prone to retrogradation. Those in which there is an increased level of amylopectin also tend to produce weaker gels. There are also turbidity differences when the starches are dispersed in water: cereal starches generally result in opaque dispersions, whereas root starches tend to give clear dispersions.

The principal sources of starch and flour used in the food industry for thickening and similar purposes are maize, wheat, rice, and potato, with sago and tapioca also having specialist uses. They each have different viscosity, gelling and retrogradation characteristics, which results in them being used in different formulations. Ranken *et al*. (1997) briefly outline some of these characteristics.

To complement the different properties of native starches, these can be physically of chemically modified in various ways:

- Pregelatinisation: the starch is cooked rapidly in water and immediately dried. This results in a powdered form of gelatinised starch which can be rehydrated in cold water. Maize and potato starch are the usual starting material for this process.

- Oxidation: starch is treated with sodium hypochlorite, peracteic acid or an alternative oxidising agent. This yields a more stable gel than can be produced with untreated starch.

Specific ingredients

- Acid thinning: treatment of starch with hydrochloric or sulphuric acid at around 60°C results in chemical modification without the starch granule swelling, and therefore avoiding gelatinisation. Sodium chloride or sodium sulphate may be used to help prevent swelling. The starch is then washed and dried. When hot or boiling water is added to the starch, it rapidly solubilises without excessive increase in viscosity, and the resultant gels on cooling are firm.

- Side chain modification and cross-linking: chemical modification with phosphoric acid or acetyl or succinic anhydride yields a range of chemically modified starches with very consistent viscosity. These starches are very useful in the production of long-life, heat-treated sauces as the heating regime and subsequent lengthy storage period has little effect on their viscosity.

With this armoury of native and modified starches, the product developer can choose the combination of characteristics that best suit the product in question, keeping in mind the way it is going to be processed and stored. Several of the non-natural derivatives are classed as additives in EU and UK legislation (see Table 3), and their presence has to be indicated by the term 'modified starch'. The 'E' number or specific name may be listed next to this term, but these are not obligatory, and are usually omitted.

Table 3 - List of EU/UK approved starch derivative additives

Starch Derivative	'E' number
Oxidised starch	1404
Monostarch phophate	1410
Distarch phosphate	1412
Phosphated distarch phosphate	1413
Acetylated distarch phosphate	1414
Acetylated starch	1420
Acetylated distarch adipate	1422
Hydroxypropyl starch	1440
Hydroxypropyl distarch phosphate	1442

Table 4 - Some typical uses of modified starches

Raw material	Modification	Typical uses
Waxy maize; Non-waxy maize	Esterification	Salad dressings; sauces and gravies; ketchup; soups; bakery fillings; dairy desserts
Waxy maize; Non-waxy maize	Etherification	Fruit fillings; dairy desserts; salad dressings; soups and sauces; processed meat
Tapioca	Etherification	Dairy desserts; noodles
Waxy maize	Emulsification (octenyl succinic anyhdride)	Salad dressings; cakes, coffee whitener and creamer
Maize, wheat	Oxidation	Meat, seafood and poultry batters
High-amylose starch	Esterification	Seafood, cheese and vegetable batters
Potato, wheat	Dextrinisation (partial hydrolysis)	French fries
Maize, waxy maize	Pregelatinisation	Baby foods; food powders; fruit fillings; salad dressings; sauces; cream fillings
Waxy maize, tapioca	Cold-water swelling	Quiche fillings; food powders; fruit fillings; salad dressing;, sauces; baby foods; microwaveable foods

Reference:

Cerestar (2001) Food starches. Brochure describing range of available products

As can from Table 4, the variety of modifications that can be carried out is considerable. The Table is a very simplistic summary of what can be carried out. Within these general areas of modification, subtle changes in the degree of modification can yield a significantly different product with markedly different characteristics and uses.

3.4 Sugar

Sugar is added to a very wide variety of products for flavouring purposes. In many cases, alternatives have been sought to the use of sucrose (table sugar) to achieve the same desired sweetening effects. These alternatives have ranged from other high-calorie bulk sweeteners, such as glucose, fructose and sugar syrups, through reduced calorie sugar alcohols (sorbitol, maltitol and xylitol) and polydextrose (available commercially as "Litesse") to the high-intensity 'artificial' sweeteners such as cyclamates, acesulfame K and aspartame (available commercially as "Nutrasweet"). The reason for using sugar alcohols, polydextrose and the high-intensity sweeteners has usually been to reduce calorie intake (high-intensity sweeteners provide no calories, and the sugar alcohols provide less than half of calories of sucrose), while still providing the same degree of sweetness. Sugar alcohols and high-intensity sweeteners also have the advantage that they are non-cariogenic (i.e. they do not promote tooth decay). Each change in formulation brings with it its own set of problems - usually related to flavour and sweetness: e.g. the sugar alcohols are less sweet than sucrose, and some of the high-intensity sweeteners leave a bitter aftertaste. There have also been questions raised as to the safety of some of the latter. In the main, many of the problems have been solved, and modifications to formulations have resulted in products that are both safe and of the desired flavour and sweetness.

In certain products, particularly cakes, however, sucrose performs many functional roles, and direct replacement with one of these alternatives is not possible. In cake batter, sucrose dissolves in the aqueous phase during mixing and helps to stabilise the foam by controlling the viscosity of the continuous phase. This means that the flour starch granules are kept suspended and prevent the air bubbles coalescing in the batter. Consequently, the final baked volume is maintained and problems such as the formation of a rubbery layer at the base of the cake are reduced (Pateras, 1991). Sucrose raises the temperature at which egg protein coagulates, and also delays the gelatinisation of starch by affecting protein and starch hydration. Both of these effects result in a more tender, softer crumb texture.

Sucrose also acts as a humectant, controlling the water activity of the cake and inhibiting mould growth. Its ability to bind water and soften the crumb also contributes to anti-staling properties. The phenomenon of staling is a complex one, and is not simply due to 'drying out' as moisture moves from the crumb to the crust of the cake. Starch crystallisation is the basis of staling: sucrose is believed to interfere with recrystallisation of starch and inhibits starch retrogradation during storage.

Finally, sucrose allows crust colour to develop at lower temperatures, by lowering the caramelisation point of the cake batter. Work at CCFRA has demonstrated that, because of the multi-functional role of sugar in cakes, a direct replacement with other sweeteners is not a simple proposition. Nine different bulk sweetening agents, including monosaccharides, disaccharides, polyols and oligo/polysaccharides were used as sucrose replacements in experimental high-ratio cake recipes (a high-ratio recipe is one in which the liquid and sugar contents are higher than the flour weight). Each of the replacements had its own effects on cake quality, and these effects varied, depending on how much of the sucrose was replaced (Anstis and Cauvain, 1998).

Sugar's role as a humectant (i.e. reducing water activity and thus acting as a preservative) is seen in many other products, particularly jams, jellies and marmalades. This subject is outside the scope of this book but is dealt with in another Key Topic ('Food preservation: an overview' - Hutton, 2004).

3.5 Salt

Salt was one of the first functional ingredients to be used by early man - in warmer climates, where preservation of food by freezing or cooling was not possible, adding salt was an effective way of storing food for relatively long periods. It would have also become apparent that salt had a significant effect on the flavour of the food, and on the characteristics of fermented food. It is unique among the ingredients discussed in this Key Topic in that it is not of biological origin.

The health impact of too much salt in the diet has been at the forefront of media and government attention in recent years. Whilst reducing the need for the flavour-enhancing aspects of salt may be seen as mainly a problem of 'educating' the

Specific ingredients

consumer, and the preservative functions can be replaced by alternative mechanisms such as chilled storage or the addition of other chemical preservatives, replacing the processing functionality of salt is a more difficult hurdle to overcome.

The multiple roles of salt were reviewed recently (Hutton, 2002a). In this Key Topic, the focus is on the processing and formulation roles of ingredients, but it is worth summarising the preservative effects of salt. Its main mechanism of action is via a reduction in water activity. Microorganisms require water to survive and grow. Salt in solution binds to some of the water, leaving a reduced level for microbial growth.

Salt can be used to preserve meat, fish, cheese, fruit and vegetable products, either as the sole preservative or in combination with acid, fermentation or both - salting of meat and fish in particular is a very old method of preserving foods. Salt can be added either in the form of dry salt, rubbed onto the surface of the food, or as a brine solution. Saturated brine is usually required to achieve long-term preservation if this is to be the only preservative factor. When the product is going to be consumed or further processed, the excess salt will usually need to be removed to yield a palatable end product.

Sensory effects

Salt not only confers its own taste on foods, it is also used to enhance and modify the flavour of other ingredients, and to reduce the sensation of bitterness in some products. Its flavour-enhancing properties are thought to be related to its effect on water activity, but its precise effects are very much product-specific. As with its preservative effects, salt 'ties up' some of the water molecules in the food. This increases the effective concentration of volatile flavour molecules, and enhances the strength of flavour. It will also effectively increase the ionic strength of the water phase of the food (the concentration of non-volatile solute molecules), which can affect the binding of proteins and other components and hence the texture (which is discussed later) and flavour (Delahunty and Piggott, 1995).

Flavour perception is a complex phenomenon, and is affected by both the chemical and physical nature of the food ingredients and their relative levels. For example, salt flavour intensity in frankfurters was enhanced by reducing the fat content from 12% to 5% (Hughes *et al,* 1998).

The level of salt added to foods for sensory purposes varies widely both within and between products. The levels used are determined by consumer demand, although it can be argued that the range of consumer tastes is also driven by what is available. Salt is added, for flavour purposes, to soups and gravies; butter and spreads, bread, biscuits, pastry and similar products, cheese, breakfast cereals, pasta and noodles, canned foods, meat products, savoury snacks, and even chocolate. In some cases, it is primarily added because a salty taste is what is desired in the final product. In savoury snacks, for example, it is usually the predominant and often the only recognisable taste, although it can also be used as a vehicle for uniformly distributing other flavour components and microingredients (e.g. vitamins and antioxidants) throughout the finished product. However, salt is also used in a more subtle way, removing 'blandness' in products such as breads, sweet biscuits, pastry and breakfast cereals, where the consumer does not specifically perceive a salty flavour. In an extensive study with soups, rice, crisps and scrambled eggs, using a trained sensory panel, salt was found to add more than just a salty flavour. It increased the impression of fullness and thickness, giving the impression of a less watery product, and enhanced the perception of sweetness (Gillette, 1985). Further examples of the sensory effects of salt in food are given in Hutton (2002).

Processing aspects

Salt is a major factor in the processing characteristics of certain food groups and in the consequent nature of the final product. Its major role in bread and other dough formation and in cheese manufacture are discussed below, but it also finds applications in a range of other food types. For example, it has a solubilizing effect on meat proteins, causing them to become sticky. This assists in the manufacture of reformed steaks and meat joints, which are popular because of their reduced level of fat compared with primary cuts. This property is also used in the production of frankfurter sausages, in which comminuted meat proteins are bound firmly together. The binding is so strong that the sausage snaps when bent.

Salt can affect the balance of biochemical reactions and can thus have an effect on all living systems. In bread dough development, the major effect, as will be seen, is on the gluten proteins of the dough. In cheese manufacture, biochemical and microbiological effects are seen.

Cheese manufacture

There are now about 500 varieties of cheese produced commercially. Despite the fact that they evolved as a result of the need to find a longer-lasting product based on milk, they are a generally not 'stable' products, and the changes that occur within the stored cheese can be manipulated to yield the variety of products that exist today. Salt is one of the ingredients that can be used to significantly affect the nature of the end product. The amount that is added and the stage in the cheese-making process at which it is added vary from one cheese to another, and this in turn affects its precise role in the development of the cheese. For example, addition of dry salt to Cheddar cheese occurs before the cheese is put into moulds (the hooping stage), and the process only takes about 15-20 minutes. In contrast, with Dutch, Swiss and Italian cheeses, immersion into brine occurs after the cheese has been removed from the mould and pressed; this can take several hours or even days.

Cheese manufacture is itself a multi-stage operation, involving enzymes and microorganisms naturally present in the milk, as well as added enzymes (rennet) to promote coagulation and microbiological cultures (see Figure 5 for a typical flow diagram of cheese manufacture). Salt can affect the activity of all of these as well as having an influence on the physical nature (e.g. structure) of the curd; clearly the combinations of level and time of addition, along with addition of all the other constituents will lead to complex matrix of possible effects. Guinee and Fox (1993) discuss in some detail the many stages of cheese manufacture and the role that salt plays in them. A few examples from here and elsewhere are given below.

Salt has a significant effect on enzyme activity. In most cheese production, initial breakdown of proteins is catalysed by enzymes remaining after the initial coagulation phase. One particular milk protein, α-s1-casein, undergoes considerable proteolysis during ripening, but another, β-casein, remains unaffected. The action of two of the enzymes involved in the proteolysis of

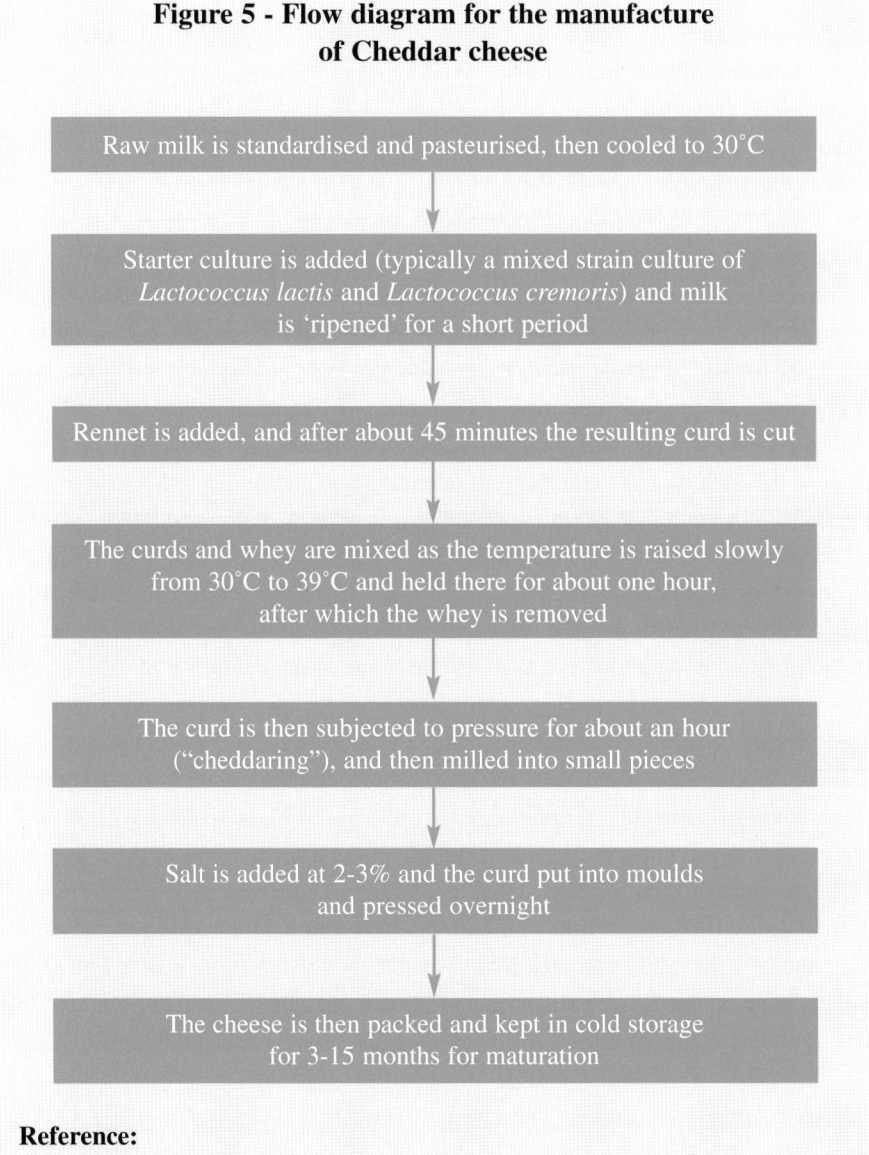

Figure 5 - Flow diagram for the manufacture of Cheddar cheese

Raw milk is standardised and pasteurised, then cooled to 30°C

↓

Starter culture is added (typically a mixed strain culture of *Lactococcus lactis* and *Lactococcus cremoris*) and milk is 'ripened' for a short period

↓

Rennet is added, and after about 45 minutes the resulting curd is cut

↓

The curds and whey are mixed as the temperature is raised slowly from 30°C to 39°C and held there for about one hour, after which the whey is removed

↓

The curd is then subjected to pressure for about an hour ("cheddaring"), and then milled into small pieces

↓

Salt is added at 2-3% and the curd put into moulds and pressed overnight

↓

The cheese is then packed and kept in cold storage for 3-15 months for maturation

Reference:

Early, R. (1998) The Technology of Dairy Products. 2nd Edition. Blackie Academic and Professional α-s1-casein, pepsin and chymosin, is optimally stimulated by 5% salt, whereas the action on β-casein is strongly inhibited.

The ripening of blue cheeses is dominated by *Penicillium roqueforti*. Germination of spores of this organism are stimulated by 1% salt, but are inhibited by 3-6% salt, depending on the variety. Subsequent growth of the organism is not so dependent on salt levels, so 1% salt is commonly added to blue cheese curd before it is put into moulds. As well as stimulating spore germination, this gives the cheese a more open structure, which helps *Penicillium* growth.

In British-style cheeses, salt plays a role in the control of the acidity. In cheeses such as Cheddar and Stilton, the desired pH has already been reached by the hooping stage, and salt is added to maintain it at that level and control microbial growth. At this stage, the starter culture organisms (predominantly *Lactococcus lactis* and *Lactococcus cremoris*) continue to ferment the lactose present. However, the level of activity is strongly dependent on the level of salt present and the salt tolerance of the individual strains present in the starter culture. Commercial lactic acid cultures are stimulated by low levels of salt, but are inhibited at levels above 2.5%. The distribution of the salt within the cheese is also important. Cheddar cheese is usually milled into particles of approximately 2cm diameter before salting and hooping, so there will be an inhibitory effect at the surface before the salt diffuses into the particles to have an effect throughout.

Bread dough development

The development of the gluten structure in bread dough, described in Section 2.3, relies to a great extent on the salt present in the mixture. Most UK and Western-style bread is made from dough containing about 2% salt by weight of flour, which gives a level of about 1.3% in the final product (Cauvain and Young, 1998). However, levels in different formulations do vary, and recently efforts have been made to reduce overall levels. Salt has a significant physical effect on the properties of wheat gluten, making it less extensible. This is often described as a binding or tightening of the dough and the overall result is a less sticky dough. Salt also affects the rate of fermentation in the dough. Too much salt is not desirable from a

Raw materials and ingredients in food processing

processing point of view - it increases the amount of energy required for adequate dough mixing, and also means that proof time has to be extended to maintain final loaf volume. However, inadequate salt levels result in excessive yeast fermentation, leading to gassy, sour doughs that result in loaves with an open grain and poor texture.

Box 18 - Salt influence on soy sauce fermentation

Recent research demonstrates how subtle yet significant the effects of salt can be on microbially fermented foods. The manufacture of soy sauce usually involves the fermentation of a soybean/wheat kernel mixture using the shoyu yeast. This is a traditional process that has been developed over many centuries. There are five different types of soy sauce available in Japan, of which Koikuchi is by far the most popular: it is also the type regularly used in the UK. Salt levels of 17-18% have traditionally been used in shoyu fermentation for the optimum development of flavour. Amazingly, it is only recently that it has been demonstrated that this level of salt is the optimum required for the production of the many individual volatile components give the sauce its characteristic overall flavour (Sasaki, 1996). Interestingly, the researchers also found that the behaviour of the yeast in a laboratory culture medium was markedly different, with the production of many of the characteristic flavour chemicals being inhibited at salt levels above 6-8%.

Reference:

Sasaki, M. (1996) Influence of sodium chloride on the levels of flavour compounds produced by the shoyu yeast. Journal of Agricultural and Food Chemistry, **44** (10): 3273-3275

3.6 Gelling agents and thickeners

The use of specific ingredients to turn a liquid or semi-liquid product into a solid or viscous product, without reducing the water content, opens up a whole new area for product development. The art is not new: gelatin to solidify meat gels and pectin for making jam have been known, albeit by default, and used for millennia. These and

Specific ingredients

many other substances with these types of properties are collectively known as hydrocolloids. Their principal characteristic is their ability to bind water and produce networks that can bind small food components. However, they are chemically very disparate, each with their own particular characteristics, so they offer a wide range of choice to the food processor, depending on the effects required. Often two or three of the hydrocolloids can be used in any particular application, either singly or in combination, and availability and cost may be the deciding factors. Of the hydrocolloids in general use, most are of plant origin. The exceptions are xanthan gum, which is a carbohydrate produced by industrial fermentation of the microorganism *Xanthomonas campestris*, gellan gum, which is an extracellular polysaccharide produced by the fermentation of *Pseudomonas elodea*, and gelatin, which is derived from animal proteins. Almost all of these gelling and thickening agents are carbohydrates; the main exception is gelatin, which is a protein. Caseins (milk proteins) may also be used to modify texture.

The carbohydrate thickeners are treated as additives in EU legislation and each has an 'E' number; a list of those permitted is given in Table 5. Although carbohydrates, they are not metabolisable in the body and provide no energy.

Table 5 - List of carbohydrate thickeners permitted in EU

Thickener	E number	Source
Alginic acid	E400	Brown seaweed
Alginates	E401-E404	Brown seaweed
Agar	E406	Seaweed
Carrageenan	E407	Red seaweed
Locust bean gum	E410	Carob tree
Guar gum	E412	Plant seeds
Tragacanth	E413	Tree exudate
Acacia gum	E414	Acacia tree
Xanthan gum	E415	Microorganism
Karaya gum	E416	Tree exudate
Tara gum	E417	Tara tree seeds
Gellan gum	E418	Microorganism
Pectin	E440	Citrus fruits, apple

They are therefore treated as dietary fibre. Gelatin and the caseins are proteins and are metabolised, and are therefore classified as nutrients rather than additives.

As well as those mentioned, other substances may cause thickening during food processing. The functional roles of starch have been described earlier, and cellulose and its derivatives are also widely used.

The properties of the individual hydrocolloids and their uses are extensively reviewed by Chapman and Baek (2002). In this section, the main properties of some of those most widely used will be briefly explained, to give a flavour of why individual combinations are suitable for specific applications, and how they can be used for roles apart from gelling. As well as their apparent physical effect on the food itself, they can also affect the feel of the food in the mouth, and the release of flavour components.

Pectin

Pectin is a polysaccharide that occurs in nearly all fruits and vegetables. The name comes from the Greek 'pectos', which means gellified or congealed. The pectin from different sources vary in chemical composition and most are either not available in sufficient quantities or do not have the correct functional properties to make them commercially viable. The two main sources from which pectin is extracted commercially are citrus peel (ideally lemon and lime) and apple pomace.

Pectin is a high-molecular-weight polysaccharide consisting of between hundreds and a thousand galacturonic acid saccharide units, some of which have methyl ester (methoxyl) groups attached.

Pectins are broadly classified as high (HM) or low (LM) in the relative number of methoxyl groups. The two groups have different setting or gelling characteristics and therefore are used in different applications. HM pectin requires a high level of soluble solids and low pH (high acidity) to gel; LM pectin requires calcium ions to gel.

Specific ingredients

Jams and jellies were the first products to use pectin, and are still the biggest use for it. The amount of pectin added to the mixture will depend on the natural pectin content of the fruit: for example, plums are higher in pectin than strawberries. In standard fruit jams (pH 3.0-3.3, total soluble solids 60-70%), a HM pectin is used. In reduced-sugar jams, the total soluble solid content is below 55%, so LM pectins are used. These require added or naturally present calcium, so components of the fruit that compete for the calcium must be considered. The way the jam is processed will also determine the choice of pectin. Jams produced by boiling in open pans will require a slow-setting pectin, to prevent the jam pre-gelling.

Pectins are also used in fruit fillings in bakery products to produce a viscous slow-flowing texture, and in confectionery products such as Turkish delight. However, because of its 'sensitivity' to the processing conditions, and the need to use strictly controlled amounts of the right blend, other hydrocolloids are often preferred.

LM pectins can be used to add texture to dairy products. By varying the type and amount added, textures can vary from firm and brittle to soft and creamy. HM pectins are used to add 'body' to reduced-calorie drinks.

Xanthan gum

Xanthan gum is a microbial polymer secreted by *Xanthomonas campestris*, a bacterium originally isolated from brassica vegetables. It has been produced and used commercially in Europe since the mid-1970s. The polymer, although variable in size, is extremely large (molecular weight between 2 and 15 million), and it is the size that gives it its unique functionality. It consists of a backbone of glucose units with alternate units having a side chain of two mannose and one glucuronic acid residues.

Even at low concentrations, it has a very high viscosity in aqueous solution; this decreases as shear forces are applied, but returns after shearing stops. It is soluble in both hot and cold water, is stable over a wide range of temperatures, salt strengths and pH values, and is resistant to enzymic degradation, meaning that it can be used in the presence of amylases, pectinases and cellulases.

Raw materials and ingredients in food processing

Xanthan gum is a thickening agent: rather than forming a true gel, it forms a pseudo-gel, which is shear-reversible. Therefore, it has a very different role to play to pectin, for example. Among its many applications are:

- Dry mixes: as xanthan hydrates rapidly in water, it thickens and gives body to the product as well as suspending any particles. It is particularly used for suspending fruit pulp in prepared fruit drinks, so preventing sedimentation.

- Syrups and toppings: xanthan provides pourability and product 'cling' when dispensed onto ice cream or pancakes.

- Dairy products: xanthan helps to control ice crystal formation in ice cream, giving slower meltdown and improving texture. In chilled dairy desserts and cream it improves viscosity, mouthfeel and long-term stability, thereby potentially extending shelf-life. In whipped products, it stabilises the air bubbles.

- Pourable dressings: xanthan suspends oil droplets in emulsions, and can also suspend particulates in herb-based dressings. It also enables the dressing to be poured easily from the bottle; product 'cling' is also an important feature in this application.

Alginates

Alginates are polysaccharides that give brown seaweed its mechanical strength and flexibility. They are made up of the sugar acids beta-mannuronic acid and alpha-guluronic acid; the polysaccharide consists of sections of alternating acids and sections of one acid or the other. The exact arrangement varies with species and also with age, more guluronic acid being added in older parts of the plant. This results in alginates with various characteristics, suited to different food applications. They can be used to form solid gels or to thicken and stabilise food systems. The gelation involves calcium ions, and once the gel is formed it is generally stable to heat.

Alginates have been used as food ingredients for over 50 years. One of the major uses of alginates is in the restructuring of food products; one example is the restructuring of fruit or vegetable pieces from pulps for pie fillings. Alginates can be used to reform the pulp into pieces of uniform size - yielding a product that has the texture that the consumer associates with the particular fruit or vegetable. In a similar way, they can be used to reform meat chunks in pet food - their stability to the temperatures used in the canning operation mean that they are particularly suited to this.

Alginates are also frequently used in bakery products. One application is the cold thickening of creams - this makes the creams stable during freezing and thawing, as well as being heat resistant.

Carrageenan

Carrageenans are polysaccharides extracted from different types of red seaweed. The polysaccharide is a linear molecule made up of galactose and 3,6-anhydrogalactose sugars, some of which are sulphated, joined in different ways. Carrageenan is a very large molecule and structurally very variable, to the extent that it is divided into three categories, reflecting its functional properties: kappa, iota and lambda. Iota-carrageenan forms elastic gels, whereas those of kappa-carrageenan are brittle. Lambda-carrageenan does not form gels. Iota and lambda are tolerant to salt and are stable to freezing and thawing. Chapman and Baek (2002) list some of the products in which the carrageenans are used.

Other polysaccharide hydrocolloids

Galactomannans

These are non-gelling long-chain polysaccharides made up of mannose and galactose sugar units; they are widely used in the food industry as thickeners. They are derived from legume seeds. Locust bean gum comes from the carob tree (*Ceratonia silqua*) and guar gum is produced by seeds of *Cyamopsis tetragonolobus*, which grows in north-west India and Pakistan. Both have high water-binding properties, forming very viscous solutions at low concentrations. Although their properties are somewhat different to those of starch, being non-caloric, they are often used in foods in place of starch, particularly in reduced-calorie foods.

Raw materials and ingredients in food processing

Gellan gum

Gellan gum is a relatively new addition to the food manufacturer's armoury. It is an extracellular polysaccharide produced by the bacterium *Pseudomonas elodea,* and is a linear polymer, consisting of a repeating unit of two glucose units, one rhamnose and one glucuronic acid. It was developed by Kelco International and produces strong, clear gels at low concentrations. Allied to its very good thermal and acid stabilities, this gives it unique properties. Although it does not have melt-in-the-mouth properties, it is being investigated as a replacement for gelatin. The significance of gelatin replacement is discussed later. A recent application demonstrates how this type of ingredient can benefit liquid products, with which a gelling agent would not normally be associated (see Box 19).

Box 19 - Gum to assist cocoa suspension in dairy beverages

The company that developed gellan gum, Gellan CP Kelco, has recently launched Kelcogel HM-B gellan gum, which is designed to provide suspension of cocoa and minerals in ready-to-drink dairy based beverages.

Chocolate-flavoured beverages have become increasingly popular, but there can be stabilization problems with these neutral pH dairy beverages. Without proper stabilization, cocoa powder deposit can form on the bottom of the container. These stability problems may also affect other drinks such as ready-to-drink coffee or tea.

Suspension of insoluble minerals and fibres in milk-based dietary and fortified beverages can also be a challenge. The company also suggests that Kelcogel HM-B gum can be used at low use levels, with no adverse impacts on flavour, and that the gum disperses without lumping, hydrates easily and is heat stable, making it compatible with existing dairy processing equipment.

Reference:

http://foodnavigator.com/news/

Specific ingredients

Gum arabic

This gum, also known as gum acacia, has been used for various purposes for about 5000 years, notably by the ancient Egyptians for mummification. It is an exudate of stems and branches of acacia trees, specifically *Acacia senegal*. The concentrations required to form a viscous solution (around 40-50%) are much higher than would be used in foods and it is often used in a emulsifying capacity rather than as a gel. Stabilisation of foam in beer is one such example. It was used in gum confectionery, but is now too expensive.

Gum tragacanth and karaya gum are also gum exudates, although with different properties to gum arabic. They are less widely used in the food industry.

Gelatin

Gelatin is derived from collagen, which is a protein component of animal tissues such as bone, skin and cartilage. Although gelatin and collagen are, uniquely, high in glycine, alanine, proline and hydroxyproline, and contain no tryptophan or cysteine, they are digestively just like any other protein, and therefore gelatin is treated, in legislation, as a nutrient and not as an additive.

Collagen is a 3-D structure comprising of cross-linked polypeptide chain triple helices. Each chain in the helix contains about 1050 amino acids. Gelatin is produced by acid hydrolysis of collagen to yield a mixture of independent polypeptide chains, and double and triple-linked chains. There are a variety of extraction processes and the degree of hydrolysis can vary markedly. The resulting individual gelatin molecules can range in size from low-molecular-weight single chains to higher multi-strand polymers with much higher molecular weights (Chapman and Baek, 2002).

The molecular weight distribution of the gelatin affects its viscosity and its gel strength, and this variability is one of the reasons why gelatin has such a range of applications. Chemical changes result from variations in the extraction process, and these too affect the properties of the final product. Specifically, if an alkaline stage is used in the extraction, more carboxyl groups are created in the molecule, its isoelectric point falls from about 9.4 to around 4.8, and consequently both its chemical and physical properties are significantly affected.

The variations that can be introduced into a specific gelatin formulation and certain key properties have, since the 1950s, made gelatin the obvious choice in many gelling and firming applications. Gelatin manufacturers point to the following as its main advantages:

- Relatively inexpensive, and only small amounts of material required
- Easily digestible - a 'natural' ingredient
- Dissolves at about 35-40°C, giving it a melt-in-the-mouth property - this also gives it good flavour-release properties
- Multifunctional - can be used in gelling, thickening, emulsifying, stabilising, foaming and film-forming
- Clarity properties - only alginate can match it for clarity, other hydrocolloids tend to be cloudier

Typical areas of food use include:

- Confectionery - in such products as wine gums and fruit pastilles. The level of gelatin used and the strength of the gelatin formulation allow products as different as soft pastilles and hard gums to be produced. Gelatin is also used in toffee and nougat-type products, as well as a whipping/foaming agent in marshmallows.

- Dairy products - in products such as yoghurt, where its neutral flavour is of use. In cultured products it can be added to the other ingredients before the introduction of the starter culture. Its water-binding properties help to prevent synerisis and separation of whey proteins. In reduced-fat products, it adds body and mouthfeel to compensate for the lack of fat. It is also used in milk-based desserts such as blancmange, as well as in mousses and soufflés.

- Low-fat spreads. Butter and margarine form very stable water-in-oil emulsions, due to their high fat content. If the fat or oil content is reduced, as in low-fat spreads, the emulsion becomes unstable. Gelatin can help stabilise the emulsion, but allow it to melt in the mouth when eaten, giving it similar eating characteristics to the full-fat product.

Specific ingredients

- Meat products. Gelatin is widely used to provide the gel around meat products. In canned ham, it fills any cavities in the meat tissue, thus giving a better appearance to the sliced product; in canned and cooked meat products, it absorbs released juices. In pâtés it both stabilises the emulsified fat and binds juices.

Its ideal suitability for these and other product types has meant that consumers have become used to the many types of products and the quality aspects of those products that have become possible because of the use of gelatin. The rise in the number of vegetarians has meant that a market has developed for several of these products to be reformulated, replacing gelatin (an animal-derived ingredient) with other non-animal gelling agents. The BSE crisis further fuelled this need. However, many products have proved difficult to reformulate, especially traditional confectionery and dessert-type products. The specific characteristics that gelatin confers on many foods have led to a finished product that cannot easily be mimicked using other ingredients.

3.7 Proteins

Proteins play a different role in food structure to most other components, due to their complex chemical structure. Proteins are made up of long chains of amino acids. Amino acids are all based on the following structure:

The chains may be hundreds of amino acids long, and as there are 20 major amino acids, there is an almost infinite number of possible protein structures that could be formed. Although the amino acids are joined linearly (there are no branched chain proteins), the chemical side groups on these chains give each protein a unique distribution of chemical charges, which results in cross-linkages being formed within a chain and, as a result, complex three-dimensional structures. These structures can be partially or wholly broken down (denaturation) to give a protein different physicochemical characteristics. In some cases, the protein can be allowed to refold to its original or new shape.

Proteins give meat its fibrous texture, and this can be mimicked to a certain extent in vegetarian meat analogue products such as soya and Quorn (see Section 4.6). Because of their potential elasticity, they can also trap gases to form a bubble

Box 20 - Replacing gelatin in chilled dairy desserts

Chilled desserts are a highly dynamic sector of the dairy market, having a 26.9% share in 2002 (Key Note, 2003); growth in the market is driven mainly by products such as mousses and traditional puddings. Gelatin is one of the major gelling agents used in this type of product, but the rise in vegetarianism, the urge to secure kosher and halal certification, and concerns left over from the BSE crisis has led to a drive by the UK food industry to offer gelatin-free products. However, the importance of gelling agents in yoghurt and chilled desserts, and the unique properties of gelatin have made this difficult to achieve. The two main characteristics that are so important are that it forms a thermo-reversible gel which melts at body temperature (conferring 'melt-in-the-mouth' characteristics which affects flavour release and, thus, overall sensory attributes), and its surface activity, which helps to stabilise aerated products like mousses.

Recent research at CCFRA has quantified and highlighted some of the problems with replacing gelatin. In studies on simple dessert systems - water jellies, milk jellies and mousses - sensory qualities were compared of products made with gelatin or pectin. Pectin mousses were found to have less air, a smoother appearance and a more gritty texture, while the gelatin mousses had a stronger dairy milk flavour. Rheological studies indicated that pectin mousses had a higher gel strength than gelatin mousses and that viscosity was correlated with set appearance and thick texture. Microscopy of the jellies showed that gelatin had a sheet-like structure, whereas that of pectin was more particulate. In the mousses, however, the gelatin structure broke with aeration.

In subsequent studies, the performance of two of the commercially available gelatin replacers were assessed in mousse formulations: one based on fats and the other on a mixture of sodium alginate, carrageenan and an emulsifier. Both were used with pectin and without gelatin, and were compared with product made with gelatin only. Among the many characteristics to be affected were firmness, colour, degree of whipping and aeration, and mouthcoating texture. The gelatin replacer formulations were found to have a higher surface activity than gelatin, which affected air incorporation.

Reference:

Chapman, S., Gilbert, C., Sahi, S. and Saraiva, C. (2005) Effect of the replacement of gelatin on the sensory properties and rheology of mousses. CCFRA R&D Report 210.

Key Note (2003). Milk and Dairy Products.

Specific ingredients

structure. This is seen in numerous applications; among those discussed in this book are gluten in bread dough (Section 2.3) and egg proteins in meringue (Section 4.2).

Box 21 - Generalized structure of an amino acid and a protein chain

$$\begin{array}{c} H \\ | \\ R-C-NH_2 \\ | \\ COOH \end{array}$$

where R is one of about 20 different chemical groupings

These individual amino acids are linked to each other through the amino (NH_2) and acid (COOH groups), thus:

```
        R1       R2
        |        |
   ---OC-C-NH OC-C-NH OC-C-NH CO-C-NH--
                          |       |
                          R3      R4
```

where R1, R2, R3 and R4 may be the same or different chemical groupings

---

3.8 Water

Water is the most commonly used ingredient in food products. It is a natural constituent of the driest of foods - even cream crackers contain 4.3% water (FSA, 2002). It is also added to most formulated foods, even those that are going to be subsequently dried, to facilitate mixing of the other ingredients and help achieve the desired final texture. Its specific role will vary widely from one product to another - a few miscellaneous examples below are used to illustrate these. Its role in ice cream texture formation is described in the next section.

Raw materials and ingredients in food processing

Added water in meat

There is a good deal of controversy at the moment over the practice of adding water to meat products - both raw and processed. The technological reason for increasing (or maintaining) the water content of meat is quite valid: it is to give the product a succulent texture and enhance its flavour characteristics. In many situations in industrial cooking and manufacturing processes, not doing this would result in a dry product that was unacceptable to the consumer. However, there is the danger that this desirable practice can be extended to the unnecessary inclusion of excess water. This increases the weight of the product (so the consumer pays extra for no extra benefit), and actually decreases the eating quality of the product. The addition of water is controlled by legislation (see Box 22).

The water is added to and maintained in the product by infusing or injecting a solution of salts and/or proteins (see Section 3.11 - polyphosphates). How much has to be added will depend on the water holding capacity or water binding capacity of the particular piece of meat. These properties depend on the properties of the individual muscle fibres in the meat and may relate to chemical and electrostatic characteristics or merely the physical arrangement of the fibres (Ranken *et al*, 1997).

Box 22 - Legislation controlling the adding of water to meat products

So that the consumer is made fully aware of where significant amounts of water have been added to certain meat products, the Meat Products Regulations 2003 (Statutory Instrument No. 2075 in England) state that the addition of water must be stated in the name of the product. The products covered are those which have the appearance of a cut, joint, slice, portion or carcase of meat (it does not apply, for example, to sausages and burgers), and where more than 5% water has been added (10% in the case of uncooked cured meats). It is also necessary to include in the name of the food the proteins or other additives that have been used to introduce or maintain this added water in the product. Thus, sliced chicken breast may be labelled as: 'sliced chicken breast with added milk proteins, phosphates and water'. Additives used for other reasons (e.g. salt for seasoning purposes), would not be included in the name of the product. However, they would appear in the list of ingredients.

Specific ingredients

Ice glaze on frozen fish products

As with addition of water to meat products, the glazing of fish products prior to freezing is a technologically sound practice, but one which must be done responsibly, so that the public does not pay for extra weight that is merely frozen water. When products such as fish fillets and steaks, or scampi, prawns and shrimps are kept in frozen storage, they can be susceptible to freezer burn. This is where the surface of the fish dehydrates. This is caused by ice crystals slowly forming on the surface of the fish, denaturing the fish proteins and bursting the cells of the fish tissue. Water is then lost from the burst cells and the fish becomes both unsightly and unpalatable. The effect is enhanced because of moisture transfer from the slightly warmer air to the very cold refrigeration surfaces. This dries the air and results in greater dehydration of the fish tissue. One way of preventing this is to glaze the fish surface, whereby water is sprayed on to the surface after freezing. This then serves as a protective coating. As well as protecting against dehydration, it also protects against oxidation reactions, which would otherwise continue, even in the frozen product, leading to discoloration and off-flavour development (such as rancidity).

In the UK, there is no definitive legislation prescribing the maximum amount of glazing permitted on these fish products. A code of practice for glazed scampi reached draft form in the 1980s, in which up to 15% by weight of glaze would have been permitted in the final product, but complications and disagreements over its natural water content meant that this was never published. However, interpretation of current legislation suggests that the manufacturer should declare the net weight of the fish core prior to glazing, as recommended by the Food Advisory Committee in 1987. LACOTS (now LACORS), the body which aids in the interpretation of the law for the benefit of enforcement officers, briefly discussed the issue in 1994, particularly in relation to prawns. They stated that the UK Association of Frozen Food Producers, the MAFF Torry Research Station and Public Analysts are generally of the opinion that about 20% ice glaze would be the maximum amount necessary to prevent oxidation and dehydration of prawns. Product weights are usually given "net of ice glaze". While the presence of an ice glaze is generally not mentioned in the name of the food, some products declare a "protective ice glaze" quite prominently near the name.

3.9 Air

Air or one its constituent components is a significant ingredient in many food products. Air addition, for example, is vital in the development of bread dough (see Section 2.3), or in the development of the correct crumb structure in cakes. It is also very significant in the production of products like soufflés, mousses, ice creams and meringues (See Section 4.2), and in a more simplistic way in the manufacture of certain aerated chocolates.

In many of its applications, the addition of air to a mixture of ingredients was experimented with to try and produce foods with improved textural qualities - the examples of mousses, soufflés and ice cream described in Section 4 demonstrate this well. In fact, any base mixture which can support a stable bubble system has the potential to use air to improve mouthfeel. In bread and cakes, the structure of the finished product depends on the inclusion of air into the mixture. In some breads, it has a second important role - the oxygen component combining with ascorbic acid (a flour improver) to develop the structure of the dough gluten. In all of these applications, the air needs a structural component in which to be trapped to form sponges and foams. Proteins are the primary structure in which these sponges and foams are formed (see Section 3.7).

Head formation in beers

In beer, nitrogen and/or carbon dioxide are used for the formation of a 'head' in draught beers. During the brewing of beers, brewers' yeast produces both ethanol and carbon dioxide, the latter dissolving in the developing beer. Extra carbon dioxide may also be incorporated into the finished keg of beer. When the beer is dispensed, the pipes to the dispenser and the dispensing tap itself cause the carbon dioxide to be released as very small bubbles. The barley proteins and associated molecules in the beer help to stabilise these bubbles so that they rise to the surface and form a head. Although the liquid surrounding the bubbles will lose both liquid and gas, and shrink, eventually causing the head to dissipate, this is a gradual process. By comparison, bubbles of carbon dioxide in a soft drink will merely rise to the surface and burst.

Specific ingredients

The length of time that the head lasts is very much a function of the beer itself, as well as the characteristics of the pouring apparatus. A beer tap, for instance, makes the best heads because it forces the beer out quickly, which results in the rapid formation of high numbers of small bubbles; smaller bubbles take longer to deflate. Stouts and other big-foaming beers often get an extra kick from an injection of nitrogen gas during the dispensing process; this makes for even smaller, longer-lasting bubbles. Nitrogen bubbles in this situation are smaller than carbon dioxide bubbles. As well as the smaller size conferring extra stability, the combination of carbon dioxide and nitrogen bubbles results in a very long lasting head. This is such a desired feature of many beers that much effort was put into trying to mimic the effect in bottled beers, and this led to the development of the 'widget' in order to incorporate an important ingredient (nitrogen gas) into the product (see Box 23).

Box 23 - The development of the widget

The widget is a good example of how the needs of the product for the incorporation of specific ingredient influenced packaging design, and spawned a new packaging development. Some bottled beers have existed for many years and have evolved or been developed into popular products with distinct characteristics of their own (e.g. Worthington 'White Shield' which is fermented in the bottle). However, many products are intended to be canned or bottled varieties of the 'draught' equivalent. One of the major problems with this was the inability of the canned or bottled product to produce a 'head', similar to that of the draught beer, when poured out. Although the beverage in the bottle is carbonated, the release of gas on opening does not mimic the effects of pulling draught beer. The problem is compounded in stouts like Guinness, which are less fizzy than lagers and some bitters (i.e. they contain less dissolved carbon dioxide, and so less is released on opening the can or bottle), and depend on nitrogen added during the dispensing process to produce a head. Nitrogen tends to form smaller bubbles than carbon dioxide and this results in a creamier, longer lasting head.

The Guinness company developed and introduced an insert (which they called a widget) in the mid to late 1980s to help achieve a 'smoother' Guinness pint. Nearly 30 alternative designs were considered before a nitrogen filler design

continued...

Raw materials and ingredients in food processing

was adopted - it was appreciated early on that removal of oxygen from the device was a key factor (Browne, 1996). Many other brewers have subsequently incorporated widget designs into their own beers.

There are now many different variations on widget design. The widget originally developed by Guinness and described in US Patent 4832968 is a small plastic hollow pod with a minute hole in it. It is placed inside the can during the first stage of the packaging process. When the can has been filled with drink, sealed and chilled, the drink inside the can becomes naturally pressurized (in some beverages, a small amount of liquid nitrogen can be included before closure to increase pressurization). This pressure forces around 1% of the drink inside the widget, in turn pressurising the gas (nitrogen) within. The widget has a small chamber inside that is specifically designed to hold a small amount of drink, which flows in through a very small opening in the underside of the widget.

When the can is opened, the contents reach normal atmospheric pressure. The drink that is held inside the chamber of the widget is forced out through the small opening in the widget as the pressure inside and outside the can equalize. The effect is to release millions of tiny bubbles of carbon dioxide from the drink, which rise to the surface and form the familiar, creamy head of a draught beer or stout. The key difference between the widget system and straightforward opening of a can is the greater number and smaller size of the bubbles released.

Widgets have been designed which are either attached to the can, or float on the surface. Variations have also been developed for use in bottles.

Reference:

Browne, J.J.C. (1996) What widget? Brewer **82** (11): 498-503

3.10 Microbial cultures

Fermenting of food using microbial cultures (or originally the naturally occurring microflora associated with food) has been used as a preservation mechanism for millennia. It evolved particularly in warm climates where it was not possible to use natural freezing or chilling facilities to extend the shelf life of food. The three major areas of microbially preserved (fermented) foods are dairy products (such as

Specific ingredients

yoghurts and cheeses), meat products, and vegetable products. In addition, both alcoholic beverages and bread products rely heavily on the action of microorganisms (although this is not strictly the preserving of a specific 'unstable' foodstuff). By adding specific 'starter' organisms, the growth of pathogens and spoilage organisms can be inhibited: the starter culture 'out-competes' them. Non-microbial spoilage - such as biochemical rotting and drying out - can also be side-stepped. In many cases, the food itself is changed so that it is intrinsically less susceptible to both microbial and chemical spoilage. For example, the microbial cultures may result in food either having a much reduced water activity or being much more acidic or alcoholic, all of which are methods of chemical preservation.

The preservative aspects of biocultures are discussed in more detail in Hutton (2004). The use of microorganisms has been further developed to produce end products with particular flavour and texture characteristics. Fermented dairy products such as cheeses and yoghurts are produced by inoculating milk with various microbial cultures. However, there are many ways in which conditions in the developing cheese can be manipulated to allow the development of different microbial populations that give a particular cheese its individual characteristics. This happens, for example, in bacterial surface-ripened cheeses, where the microbial colonies develop on the cheese and are smeared over the whole surface. The composition of this smear depends on which microorganisms are present in the brine used in curing and in the factory itself, which in turn is related to the humidity and temperature of the atmosphere, the degree of ventilation, and the overall use of the building (Fox, 1993b). Similarly, varying the microbial contents of meat fermentations by altering salt content and water activity can lead to a variety of different products.

One further example of interest is the production of *Botrytis* wines. *Botrytis cinerea* is a fungus which, given the right conditions, develops on grape bunches. It is also known as 'noble rot'. Conditions for *Botrytis* growth have to be ideal as there is a fine line between noble rot, which can double the sugars to twice their normal levels, and grey rot, which can cause the fruit to become unusable. The *Botrytis* spores leach moisture from the berries, causing the fruit to break down; this concentrates the sugars and allows the production of a wine of intense richness. Among the many sweet wines that rely on *Botrytis* for their production are Sauternes and various Reislings. Growing conditions may not always be right for the development of *Botrytis*, and changes to the growing regime may need to be made to ensure that the grapes harvested are suitable for production of wine with the desired characteristics.

3.11 Other 'additives'

A variety of other materials are regularly added to formulated foods for many different purposes. Preservatives and antioxidants serve to prolong the shelf-life of the foods from a microbiological and chemical perspective. Acidulents, salt, and sugar alter the chemical properties of the food, with similar consequences. However, there are additives other than those already discussed that affect the structure and subsequent eating quality of the food. A selection of these are described below.

Calcium

The addition of calcium salts to canned vegetables is a relatively common method for improving the texture of the final product. Many vegetables when canned may become soft and even mushy. This is because the combination of heat and water causes breakdown of the tissues of the vegetable that give it its rigidity. The addition of calcium salts helps to preserve that rigidity. Dipping peeled tomatoes in solutions of calcium chloride causes the formation of a calcium pectate gel that supports the tissues and minimises softening during heat processing. It was calculated that increasing the calcium content by 100-300ppm (equated to dipping in 2% solutions for 2-3 minutes), would cause a desirable increase in firmness. As an alternative means of adding the calcium, calcium chloride tablets may be added during the can filling process (Kilcast, 2004). The levels that can be added are limited by the fact that calcium chloride at higher levels is associated with a bitter flavour in the product. In potatoes, which can disintegrate on processing, calcium lactate has been used as an alternative, but this is more expensive.

Specific ingredients

Phosphates

Phosphates in various forms are used for different purposes in a wide variety of formulated foods. The chemistry of the phosphates is complex and the nomenclature surrounding the group can also be quite confusing! It is a trivalent anion (PO_4^{3-}) which can bind with different combinations of cations (e.g trisodium phosphate or monopotassium dihydrogen phosphate). Molecules with two or three phosphates joined together (the pyro- or diphosphates and the tri- and polyphosphates) also exist, and each of these has its own specific characteristics and roles in various products. For example, the monophosphates (usually called orthophosphates) are used as emulsifiers in cheese. Acid calcium phosphate and acid sodium pyrophosphate are used as the acidulent in baking powders. Tripolyphosphates are added to meat products because of their water-binding properties.

The mode of action of the polyphosphates in meat is quite interesting. All inorganic salts have some influence on the water-binding and related properties of meat, and the degree of this effect is usually related to the ionic strength of the salt in solution (Ranken *et al*, 1997). However, pyrophosphates and polyphosphates seem to have a synergistic effect with common salt (sodium chloride) increasing the latter's influence on water binding, cooking losses, meat and fat binding and texture. The result is a more succulent product, which loses less water and oil (and associated water-soluble and fat-soluble flavour compounds) on cooking. There is a limit to how much phosphate can be added. About 0.3% is the norm (although a§ little more is sometimes used if there is likely to be uneven distribution in the product); above this level, a bitter taste becomes apparent.

Raising agents

There are several different combinations of chemicals that can be used as raising agents for the aeration of bakery products. Baking powders are mixtures of sodium bicarbonate and an acid which, in the presence of water, will react to produce carbon dioxide. The two components are usually dispersed in an inert carrier such as starch. Sodium bicarbonate will produce carbon dioxide without any acid present, when heated, but the residue is very alkaline (basically it is sodium hydroxide), which produces an unpleasant flavour and causes cake crumb to turn yellow (Ranken *et al*, 1997). One of several acids is added to the bicarbonate, so that the final product, after the release of carbon dioxide, is just on the acid side of neutral. The choice of acid will depend on flavour constraints and speed of action required. The four principal acid components used by industry are acid calcium phosphate, acid sodium pyrophosphate, cream of tartar (potassium hydrogen tartrate) and glucono-delta-lactone. Cream of tartar was the traditional acid component for baking powders, but has the disadvantage of causing rapid release of carbon dioxide and has been largely replaced by the acid phosphates. However, the latter tend not to result in such a satisfactory flavour in the final product. Glucono-delta-lactone does not have this acidic bite, and also results in slow release of carbon dioxide. However, it is more expensive than the other acidulents.

Ascorbic acid

Ascorbic acid (vitamin C) exerts its role in the body and in many food products by its antioxidant activity. The primary role of antioxidants in foods is to prolong the shelf-life of the product by preventing chemical deterioration of the food such as off-flavour development or adverse colour changes (e.g. rancidity in meat and fish products, or browning of cut fruits and vegetables). This role is discussed in more detail in Hutton (2004). However, vitamin C can also be used in a functional role - in particular in some bread making processes. In this particular case, it is the oxidising power of dehydroascorbic acid that is significant.

Basically, ascorbic acid works by 'mopping up' oxidising potential around it, being converted to dehydroascorbic acid, and preventing other chemicals (in the body or in food) from being oxidised. In breadmaking, it is used to improve dough gas retention through its effect on gluten structure. During dough mixing, the oxygen incorporated into the dough readily oxidises the ascorbic acid to dehydroascorbic acid, catalysed by the enzyme ascorbic oxidase, which occurs naturally in wheat flour. This then oxidises the sulphydryl (-S-H) groups of the gluten-forming proteins to yield disulphide (-S-S-) bridges in the proteins. The net effect is that the ability of the dough to retain gas is improved, and final bread has a finer crumb cell structure. From the consumer's point of view, the crumb is softer to the touch, but is resilient to compression (i.e. it springs back to its original shape) (Cauvain and Young, 2001).

Specific ingredients

Because the role of ascorbic acid is dependent on the mixing of oxygen into the dough, its effectiveness varies from one production technique to another. It also means that its effect is mainly limited to the dough mixing period, because the bakers' yeast in the dough will subsequently remove any remaining oxygen from the air bubbles. It also means that, in some processes which involve extended fermentation periods, it may start to reverse its reaction, weakening the gluten.

Ascorbic acid is the only oxidant improver allowed for use in bread making in the EU, although potassium bromate used to be permitted and is still commonly used in some parts of the world.

Bulking agents

These are low-calorie or calorie-free fillers that can be used to replace or supplement other ingredients in a range of low-calorie and similar products. The major non-calorific bulking agents are derived from cellulose and are insoluble in water. The main low-calorie fillers are polydextrose and the polyols (sugar alcohols), which both provide about one quarter of the calories of the same mass of carbohydrate.

Sugar alcohols can be used as sweeteners to replace sucrose or glucose, but they can also be used in much higher volumes as a bulk filler.

4. EXAMPLES OF THE USE OF INGREDIENTS FOR A SPECIFIC FUNCTIONAL PURPOSE

4.1 Laminated fat products

In previous sections, the importance of fats and flour proteins in different types of products has been discussed. In laminated pastry products such as puff pastry and the yeasted Danish pastries and croissants, it is the physical structure of the dough and fat components (i.e. that they are present in multiple thin layers), as well as the physicochemical properties of the ingredients, that give the final product its characteristics. As ever, ingredients with specific characteristics are suited to different products.

The structure of these products is based upon extremely thin layers of dough and fat - with approximately 70-100 layers of fat being present in puff pastry. There are several techniques by which this is achieved - commonly known as the English, French and Scotch methods. In the English method, an initial structure of two fat layers and three dough layers is produced by covering two thirds of a rectangular piece of dough with the laminating fat; the uncovered dough is then folded over half of the fat (see Figure 6). The uncovered half of the fat, together with the dough underneath, is then folded over the first fold to give three dough layers and two fat layers.

The thickness of the layers is reduced, and their number is increased by either multiple folding operations or by cutting the dough and stacking several layers on top of each other (Cauvain, 2001).

There are many factors that will affect the quality of the final product: the factors that are most important will vary according to product. For example, pastry lift can be increased by adding low levels of hydrocolloid, probably because this improves the gas-holding capacity of the dough. However, maintaining correct dough

Examples of ingredients and function

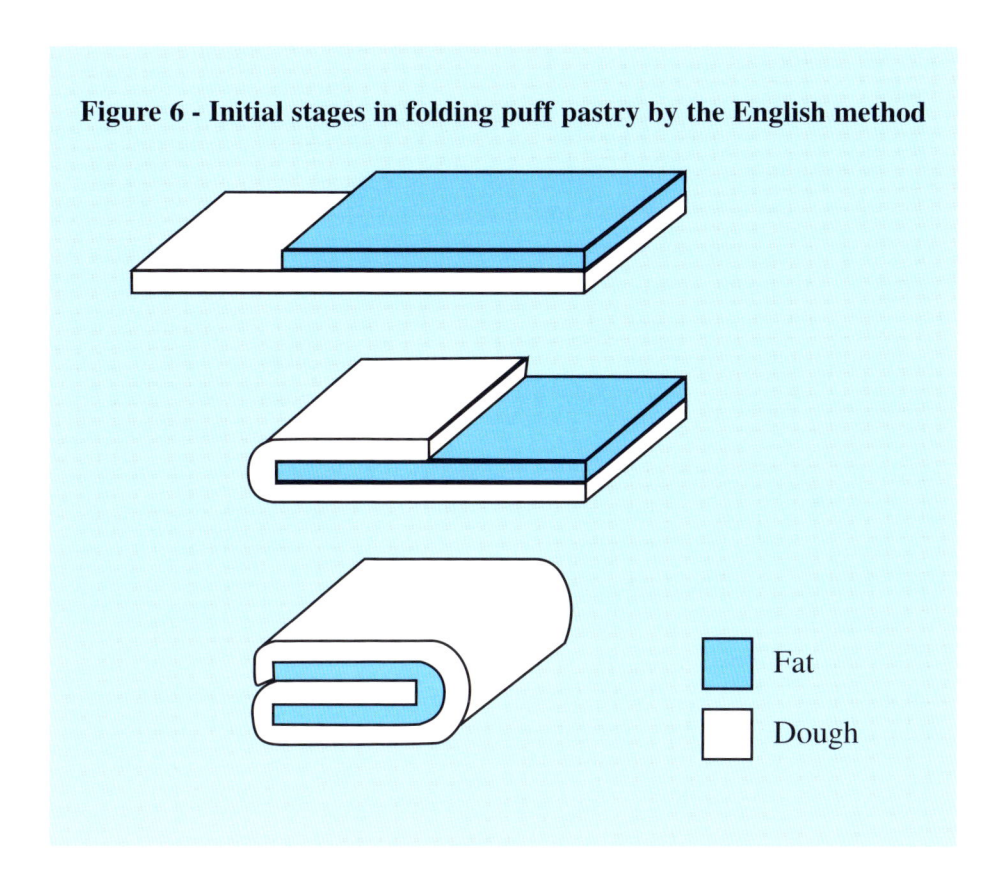

Figure 6 - Initial stages in folding puff pastry by the English method

consistency is more important for fermented laminated products such as croissants than for puff pastry.

Changes in the quality or amount of laminating fat in puff pastry structures causes significant changes in puff pastry lift and eating characteristics. Of particular importance is the softness (plasticity) of the fat; this is dependent on the crystalline solid fat content and crystal size, both of which are strongly influenced by storage and processing temperatures. Plasticity is increased by input of work: physically working the fat to make it softer. If laminating fats are not plasticized they become brittle and tend to break into lumps when first deformed. This leads to irregular formation of layers in the paste, and increased variation in pastry lift, shrinkage and general quality (Cauvain, 2001).

4.2 Meringues

Meringues usually contain only 4 basic ingredients: egg white, sugar, an acid component and air, although additional flavourings such as vanilla or salt can be incorporated. The key factor is the maintenance of a bubble foam structure by the egg albumens. Beating an egg white traps air within it. This creates a foam and increases the egg white's volume by six to eight times so that it stands in peaks with the help of stabilizers, such as cream of tartar (which is acidic) and sugar. Heating the foam by baking causes the tiny air cells to expand and the egg protein to coagulate around them, thus forming the final expanded, set meringue.

Before baking, the foam can easily be destroyed by ingredients that destabilise its structure: oils or fats (e.g. from egg yolk) do this very readily. It has been suggested that, because plastic mixing bowls are more susceptible to containing oil residues after cleaning, better meringues can be made using glass or metal bowls. The degree of beating of the egg white and the acid component (if included) is significant: Underbeating reduces the volume. Overbeating makes the whites stiff and brittle, and this makes it harder to blend other ingredients into them. An overbeaten egg white also does not expand properly when heated.

So what exactly is happening during the process? Whisking causes the protein in the egg whites to unfold - basically, their natural three-dimensional structure is partially destroyed. The addition of an acid also causes structural changes in the proteins. The resulting denatured proteins form films that trap the air bubbles; the addition of sugar stiffens the foam, by lowering the amount of available water in the protein mix (the sugar basically 'occupies' some of the water molecules). By varying the amount of sugar in the final mix, the degree of hardness or softness in the final product can be controlled. Anecdotal evidence suggests that leaving the egg white to acclimatise to room temperature before whisking improves foam formation. It is also suggested that very fresh eggs do not foam well, and that protein changes occurring in the egg white 5-6 days after laying are beneficial to meringue production (Anon, 2004). As the mixing time increases, the bubbles become smaller and more numerous; this increases the volume and makes a more stable structure. However, overbeating will cause the proteins to lose their ability to hold the small air bubbles - the albumen becomes 'overstretched' - causing the meringue to lose volume or collapse. Finally, cooking (baking) a meringue will set the structure. During heating, the air bubbles expand, causing the volume to expand until the white's protein network surrounding the bubbles solidifies. It is also important that the meringue has not been over-beaten at this stage, as it will be unable to expand further as the bubbles increase in size (Stadelman and Cotterill, 1973).

4.3 Soufflés

Soufflés rely mainly on egg white foams for their 'lift' and light structure. The word actually means 'blown'. They can vary enormously in their style - basically they are formed by folding beaten egg whites into a flavoured base. This can be either sweet or savoury and can vary widely in composition. The nature of these ingredients will affect how the soufflé sets. It is important to incorporate the egg white meringue structure carefully, as it is easy to destroy the small-bubbled foam structure. When the soufflé is baked, the heat of the oven causes the trapped air in the egg whites to expand, as in meringue formation described above. During this process, steam also assists in the mechanical leavening and the soufflé pushes up. Once the mixture reaches its maximum expansion, proteins in the ingredients begin to coagulate, and any starches present gelatinize, setting the structure. The main difference in the setting of the structure in soufflés and meringues is that, in the latter, the egg white foam becomes dried out and a crisp final product is formed with a more or less rigid structure. In soufflés, although the heat coagulates the egg white proteins, there is much less drying out of the product in total. As some of the 'lift' is due to steam generation and expansion during the baking, once it is removed from the oven it begins to subside.

4.4 Ice cream

Ice cream is a product in which fat, water and air are combined in a unique way to give a very distinctive textured product. To combine these in a stable way and incorporate different flavours is a complex operation that has evolved significantly since the first 'cream ice' became available in the UK in the mid-1700s. Ice cream provides a very good example of how processing technology can produce a very acceptable product from simple starting ingredients, and also how enforced changes in both the major fat ingredient and many minor ingredients completely changed the formulations that were used. Crowhurst (1993) gives a fascinating and highly readable account of the history of ice cream development; below is a summary of some of the points.

Early ice cream recipes usually comprised milk, eggs and cream, but between 1860 and 1875, a large number of Italian immigrants to the UK began to develop the product. In the USA, two significant technological developments, the mechanically refrigerated freezer and the homogeniser, both complemented and affected recipe developments in the UK, two of which were the addition of arrowroot for thickness and stability, and cornflour for texture. Gelatin also became widely used as a stabiliser. As ice cream is a frozen product, the development of its formulation has been significantly affected by freezing technology, and particularly the availability of electricity to power both freezers and production lines. The other major effect on ice cream formulation and the use of different ingredients, in the UK at least, was the Second World War: in 1941, a total ban on ice cream production came into force, as resources had to be diverted to the war effort.

After the war, when there was a sudden boom in ice cream production, a wide range of substitute ingredients had to be used, since the normal range of materials were in short supply. In particular, vegetable fats and oils were used to replace milk, cream and butter, which were still unavailable. Also in short supply were sugar and cornflour - other flours such as soya flour were used instead of the latter. Trade in gelatin was also banned at this time, and sodium alginate was used in its place. As has been explained in earlier sections, replacing one gelling agent with another will alter the characteristics of a product, and will require other reformulations to balance these changes. The ice cream industry was faced with the task of producing a product that was as acceptable as the pre-war product, and as similar in eating quality to it, from almost a completely different set of ingredients. That they succeeded so well has led to its own problems, with differences over what products the term 'ice cream' can be applied to.

In the USA, there are legislative standards for ice cream which require it to be a dairy product. It is described as: 'a food produced by freezing, while stirring, which is a pasteurised mix containing at least 10% milk fat, 20% total milk solids, safe and suitable sweeteners, and defined optional stabilising, flavouring and dairy-derived ingredients'. Other food fats are excluded except as components of flavouring ingredients or in incidental amounts added for functional purposes (Marshall *et al*, 2003). In the UK, after much debate, it was decided in 1959 that the term ice cream had become so synonymous with the non-dairy as well as the dairy product, that it should continue to be allowed to describe the product. By this time, although dairy ingredients had become widely available again, the public had accepted the non-dairy version, which could be produced significantly more cheaply. However, the concept of 'dairy ice cream' was also introduced to distinguish the two products.

The initial mix of a typical ice cream recipe consists of the fat (milk-derived or a vegetable fat) and the aqueous phase, in which the sugar, non-fat milk solids, emulsifying agents and stabilisers are completely or partially dissolved. An emulsion of the two phases is produced by passing the entire mix through a mechanical pump which forces the mix through fine valves at very high pressure. This breaks the fat into very small globules of uniform size, and the surface tension of these globules prevents them from coalescing. The emulsifiers in the recipe help to make this emulsion permanent.

The key stage and ingredient in the subsequent production of ice cream is the incorporation of air during the freezing process. This increases the total volume significantly from that of the initial mixture, and this volume increase is called 'over-run', and is vital in obtaining a good texture. The degree of over-run has to be controlled as, if excessive, it can result in poor quality as well as the consumer being duped into thinking they have bought a greater weight of ice cream. The volume increase is significant: up to a doubling of volume can be achieved. In the US, legislation effectively caps the permitted amount of over-run at this level. In the UK there is no such legislation, although a proposal to require the weight of ice cream in a pack to be declared was put forward in 1990 (Crowhurst, 1993).

Although the lack of controlling legislation in the UK regarding increasing the volume of the ice cream through incorporation of air might seem to invite

Raw materials and ingredients in food processing

malpractice, the degree of over-run that can be introduced is self limiting. The addition of air to an emulsion results in a three-phase system which contains air/fat surface interfaces, as well as fat/water interfaces. This type of system is called a foam, and is potentially unstable. The balance of ingredients used is very important. Two significant factors are the homogeneity of the original emulsion and the presence of a sufficient quantity of total solids in the mix. Thus, incorporating too much air will weaken the foam and drastically reduce stability and overall quality.

Box 24 - What is a mousse?

Mousse is a word used to describe a range of products with a light, fluffy texture. Products can be as different as a milk pudding and a savoury salmon hors d'ouvre. As with a soufflé, the base can be almost anything, but the light texture is achieved by folding in both whisked egg whites and whipped cream, the latter being an aerated product in which both milk fat and milk proteins play a role. Mousses can be stabilised by the incorporation of gelatin or a similar functional ingredient into the base mix, or may be frozen after moulding.

4.5 Mayonnaise

Mayonnaise is an emulsion of vegetable oil, egg yolk or whole egg, vinegar, lemon juice, salt and other seasonings such as mustard. The oil content is usually 70-85%; it is difficult to produce a quality product with less than this - the stiff body of the final product being dependent on fat content unless the product is completely reformulated (see Box 25). The key phase of mayonnaise production is formation of an emulsion containing the correct sized oil droplets. The oil forms a discontinuous internal phase in the external aqueous phase. In a good mayonnaise, the largest droplets are no more than 8µm in size and many are in the range 2-4µm. The temperature of mixing is important in attaining this and achieving a good, thick emulsion. Above about 16-21°C, the product becomes too thin. It is important to use oils that do not crystallise at refrigerator temperatures, as this will break down the emulsion; the oil and water phases will then separate when the product is warmed up. Also important are the egg and mustard, which provide the emulsification capability, and the correct levels of vinegar and salt, for preservation purposes.

Salad cream is a cheaper product than mayonnaise, in which oil content may be as low as 30%. It may contain correspondingly less emulsifier than mayonnaise, but additional thickening agent is required to increase the viscosity of the water phase. Rather than being a stiff, but light, product, it is a thick, viscous pouring sauce.

Box 25 - Formulation of reduced-fat mayonnaise

Mayonnaise is a very-high-fat and therefore very-high-calorie product. It is clearly a product where a reduction in the oil content would have a significant impact on the energy density of the product. However, as indicated in the main text, mayonnaise relies for its texture on the high oil content, so mimicking this is a significant challenge. There is no single 'correct' way of achieving this, but the many products on the market use a combination of starch derivatives and thickeners to achieve the light but stiff texture of mayonnaise. One formulation used a mixture of tapioca dextrin, modified starch and xanthan gum. Combinations of starches, maltodextrins, gums and polydextrose are typically employed to achieve the desired consistency with a much lower oil content and a significantly reduced energy content. Even potato fibre has been used, and it is possible to produce a product with mayonnaise characteristics with no added oil at all. A typical commercial recipe for reduced-fat mayonnaise is given below:

Spoonable low-fat mayonnaise

Water	50.2%
Soya oil	20%
Glucose	5%
Vinegar (5% acid)	12%
Sucrose	3%
Corn starch	4.1%
Homogenised egg	2%
Salt	1%
Whey protein	2.3%
Guar gum	0.1%
Methocel K	0.3%

4.6 Texturised vegetable protein (TVP)

Mankind has a history of eating both animal and plant-based foods. Although the consumption of meat was originally driven by the need for the nutrients it provided (in particular, the high protein content and the high nutritional value of that protein), this was supplemented over time by a liking for the flavour and texture. In modern Western societies there has been a growing proportion of people who have wanted to reduce (or eliminate) their intake of meat - either for moral reasons or for health reasons. Although this can be achieved merely by replacing meat with completely different products, the constraints of modern living and the existing patterns of eating and available foods meant that a market developed in the late 1960s and early 1970s for replacing meat in products with analogous ingredients. There were two main requirements for the new ingredients: they would need to have a similar eating quality to meat and they would have to be nutritionally acceptable. The nutritional aspects are discussed in Hutton (2002b) - it is likely that, had development been starting now from scratch, the final nutritional profile may have been slightly different. Product texture has also been modified over 35 years, in light of new technological developments and in response to changing consumer preferences.

In 1970, a meat-like soy protein product was marketed by Archer Daniels Midland under the trade name TVP (textured vegetable protein). It is produced from soy flour and concentrated by thermoplastic extrusion to impart a meat-like texture to the products. This is achieved by mixing with water and additives to form a dough and extruding under high temperature and pressure to obtain the fibrous texture. A spinning process can also be used in which a soy isolate is solubilised in alkali and forced through a spinnerette into an acid bath to coagulate the proteins. The fibres formed can be combined into bundles. The use of the extrusion technology led to continuous slabs of texturised proteins suitable for cutting into various shapes and sizes, resulting in ham, beef and poultry-like products. As the proteins are produced from a defatted soy flour, they are themselves low in fats. However, final products often have significant amounts of fat added back in to give succulence and other eating quality attributes to the product. Ironically, there has been some criticism in the press than these vegetarian alternatives are as high in fat as their meat counterparts (which is contrary to popular misconception that vegetarian products are automatically low in fat). One might conclude, therefore, that the product

Examples of ingredients and function

developers had been successful in achieving their original goal of producing a series of vegetarian alternatives that have the functional and eating qualities of meat.

Recently, the technology has been taken one step further. ADM has developed a new range of soy products. According to ADM, the NutriSoy Next™ meat alternatives are high-moisture protein products with a taste and 'whole muscle' texture (Food Navigator web page Feb 2005). These new products are an example of development being driven by the combined forces of nutritional/ethical requirements and the necessity to improved the eating quality of the food. Soy ingredients are no longer the domain of specialist soy food companies, but are increasingly being taken up by firms looking to diversify a product range or add a health element. ADM said that, in producing the NutriSoy Next range, the firm had combined soya proteins with other vegetable proteins 'for a succulent effect.'

Box 26 - The development of Quorn mycoprotein

Quorn is now utilised in a very similar way to TVP - providing meat-like texture to non-meat products, and potentially creating a vegetarian meat analogue product. However, it was not developed specifically with this in mind.

Quorn is the registered trademark of a mycoprotein-based family of products. In the 1960s, Ranks Hovis McDougall (RHM) put a great deal of effort into identifying a 'new' food source to try and help address a potential world food shortage that was envisaged by many at the time. A fungus, *Fusarium graminearum*, was identified near wheat fields in Marlow and showed promise as a significant source of protein (mycoprotein = fungal protein). Work concentrated on determining whether it would be safe for human consumption and, in a joint venture with ICI, whether it could be produced via fermentation technology in large enough quantities and cheaply enough to be commercially viable (Wilson, 2001).

In 1984, MAFF granted approval for its use as a human food (Sadler, 1988) and the first Quorn product - Savoury Pie - was launched through J. Sainsbury in the following year. In the late 1980s, a joint venture between RHM and ICI was created (Marlow Foods) to commercially develop the new food. Products were

continued...

introduced into Belgium in 1991 and the Netherlands and Germany in 1992 (Sharp, 1994), but it was not until 1993 that products became available nationally in the UK and across Western Europe.

The acceptance and development of Quorn from a consumer's point of view (in addition to the technical and commercial feasibility issues mentioned above) has depended on two main characteristics: its good nutritional profile, and the ability to produce foods with a good flavour and texture profile from it. The texture can mimic red meat mince or poultry pieces and fillets. Quorn is relatively high in protein (12.2%) and fibre (5%), and low in total and saturated fat (2.9 and 0.6% respectively). The protein has a high biological value - i.e. it has close to the ideal combination of amino acid levels for our dietary needs (similar to the milk protein, casein), the limiting amino acids being methionine and cysteine. Fibre levels are higher than in most fresh vegetables. Quorn also contains most of the B vitamins, with the exception of B12, which is absent.

In the formulation of products, the Quorn is mixed with egg albumen and flavours and processed into the desired texture (different processes will result in different types of texture to suit the end product). As Quorn has little flavour itself, other desired flavours can be added, again to suit the end product.

References:

Sadler, M.J. (1988). Quorn. Nutrition and Food Science, **112**: 9-11

Sharp, T.M. (1994). Development and significance of a novel food. Quorn mycoprotein. Voeding, **55** (1): 28-29

Wilson, D. (2001). Marketing mycoprotein. The Quorn foods story. Food Technology, **55** (7): 48,50

5. CONCLUSIONS

The use of novel combinations of food raw materials to produce end products with improved or merely different characteristics has been practised since very early in mankind's history. Starchy flour from wheat or rice, for example, was used to thicken stew and soup-type meals, and the physicochemical properties of the flour itself were used as a base for the development of novel end products - bread from wheat flour probably being the most widespread example. Blending different raw materials together allowed a wide range of new products to be developed. Improved understanding of the functional characteristics of these raw materials (e.g. the proteins in egg white, or the emulsifiers in egg yolk) allowed more complex and varied products to be formulated, and this has evolved into the processed foods industry that we know today.

Nowadays, raw material producers plant crops with specific end uses in mind (e.g. fruit and vegetables for the fresh market, or for frozen or canned products; wheat for bread, biscuit or cake making), and whole industries have grown up around the production of ingredients for formulated foods (e.g. emulsifiers, fats, starches, and gelling agents). Whereas these ingredients are sometimes viewed as slightly undesirable, they are only extrapolations of the chemicals that occur in food raw materials (such as starch from flour, emulsifiers from eggs, gelling agents from fruit).

6. REFERENCES

The items in bold were of particular use in the production of this book, and the reader is referred to these for further information.

Adams, J.B. (2004) Raw materials quality and the texture of processed vegetables. In: Kilcast, D.E. Texture in Foods. Volume 2. Solid Foods. Woodhead Publishing, pp343-363

Anon (2004) Whisking disaster. New Scientist, 4 Dec, p89.

Anstis, J. and Cauvain, S.P. (1998) The effects of sugars and alternatives in cakes. CCFRA R&D Report No. 68

Banks, W. and Greenwood, C.T. (1975) Starch and its components. Edinburgh University Press

Bhattacharjee, P., Singhal, R.S. and Kulkarni, P.R. (2002) Basmati rice: a review. International Journal of Food Science and Technology, **37** (1): 1-12

Brown, J. (1985) (ed.) The Master Bakers' Book of Breadmaking. 2nd Edition. The National Association of Master Bakers, Confectioners and Caterers.

Browne, J.J.C. (1996) What widget? Brewer, **82** (11): 498-503

Caperuto L.C., Amaya-Farfan J., Camargo C.R.O. (2001) Performance of quinoa (*Chenopodium quinoa* Willd) flour in the manufacture of gluten-free spaghetti. Journal of the Science of Food and Agriculture, **8** (1): 95-101

Catterrall, P.F. (2000) The production of cakes from non-chlorinated cake flour. CCFRA Review No. 20

Catterrall, P.F. (2001) Fat replacers and substitutes. Challenges in the development of reduced fat bakery products. In: 'New Technologies - The Future Today'. CCFRA Symposium Proceedings

Cauvain, S.P. (2001) The production of laminated bakery products. CCFRA Review No. 25

References

Cauvain, S.P., Hodge, D.G. and Screen, A.E. (1988) Changes in the fat component of cakes and biscuits to meet dietary goals. Part 1. Fat reduction in cakes. FMBRA Research Report No. 140

Cauvain, S.P. and Young, L.S. (1998) Technology of Breadmaking. Blackie Academic and Professional

Cauvain, S.P. and Young, L.S. (2001) Baking Problems Solved. Woodhead Publishing

Chamberlain, N., Collins, T.H. and Elton, G.A.H. (1961) The Chorleywood Bread Process, BBIRA Research Report No. 59

Chamberlain, N., Collins, T.H. and Elton, G.A.H. (1962) The Chorleywood Bread Process. Bakers Digest, **36** (Oct): 52-53

Chapman, S. and Baek, I. (2002) Gums and thickeners: a review of food hydrocolloids. CCFRA Review 34

Chauhan, S., Lindsay, D., Rey, M.E.C., and von Holy, A. (2001) Microbial ecology of muffins baked from cassava and other nonwheat flours. Microbios, **105** (410): 15-27

Cook, S. (2002) Factors affecting the production of cakes from heat-treated cake flour. CCFRA Review No. 20 Supplement 1

Cox, A.E. (1967) The Potato. A Scientific and Practical Guide. W.H. and L. Collingridge

Crowhurst, B. (1993) Manual of Ice Cream. J.G. Kennedy

Dick, J.W. and Matsuo, R.R. (1988) Durum wheat and pasta products. In: Wheat - Chemistry and Technology. Vol. 2. 3rd Edition. (Ed.: Y. Pomeranz). American Association of Cereal Chemists. pp507-547

Dean, T. (2000) Food intolerance and the food industry. Woodhead Publishing.

Delahunty, C.M. and Piggott, J.R. (1995) Current methods to evaluate contribution and interaction of components to flavour of solid foods using hard cheese as an example. International Journal of Food Science and Technology, **30** (5): 555-570

Downing, D.L. (Ed.) (1988) Processed Apple Products. AVI Publishing

Early, R. (1998) The Technology of Dairy Products. 2nd Edition. Blackie Academic and Professional

Fast, R.B. and Caldwell, E.F. (2000) Breakfast cereals and how they are made. 2nd Edition. American Association of Cereal Chemists

Food Advisory Committee (1987) Food Advisory Committee Report on Coated and Ice-Glazed Fish Products. Ministry of Agriculture, Fisheries and Food. Report FdAC/Rep/3

Food Standards Agency (FSA) (2002) McCance and Widdowson's The Composition of Foods. 6th Edition. Royal Society of Chemistry

Fox, P.F. (1993a) Cheese: Chemistry, Physics and Microbiology. Volume 1. General Aspects. Chapman & Hall

Fox, P.F. (1993b) Cheese: Chemistry, Physics and Microbiology. Volume 2. Major Cheese Groups. Chapman & Hall

Galliard, T. (Ed) (1987) Starch: Properties and Potential. Critical Reports on Applied Chemistry, Vol. 13. John Wiley

Gillette, M. (1985) Flavour effects of sodium chloride. Food Technology **39** (6): 47-52

Guinee, T.P. and Fox, P.F. (1993) Salt in cheese: physical, chemical and biological aspects. In: Cheese: Chemistry, Physics and Microbiology. Vol 1. (Ed P.F. Fox). Elsevier Applied Science.

Holland, B., Unwin, I.D and Buss, D.H. (1988) Cereal and Cereal Products. Third supplement to McCance & Widdowsons The Composition of Foods. 4th Edition. Royal Society of Chemistry

Hughes, E., Mullen, A.M. and Troy, D.J. (1998) Effects of fat level, tapioca starch and whey protein on frankfurters formulated with 5% and 12% fat. Meat Science, **48** (1-2): 169-180

Hutton, T.C. (2004) Food preservation: an overview. Key Topics in Food Science and Technology No 9. CCFRA

References

Hutton, T.C. (2002a) Technological functions of salt in the manufacturing of food and drink products. British Food Journal, **104** (2): 126-152

Hutton, T.C. (2002b) Food chemical composition: dietary significance in food manufacturing. Key Topics in Food Science and Technology No 6. CCFRA

Irvine, G.N. (1978) Durum wheat and paste products. In: Wheat - Chemistry and Technology. 2nd edition Ed.: Y. Pomeranz. American Association of Cereal Chemists

Juliano, B.O. (1985) Rice: Chemistry and Technology. 2nd Edition. American Association of Cereal Chemists

Key Note (2003) Milk and Dairy Products

Kilcast, D.E. (2004) Texture in Foods. Volume 2. Solid Foods. Woodhead Publishing

Lisinska, G. and Leszczynski, W. (1989) Potato Science and Technology. Elsevier Applied Science

McEwan, J.A. (1999) Barriers to the consumption of reduced fat bakery products: a consumer approach. CCFRA R&D Report 80

McEwan, J.A. and Clayton, D. (1999) Barriers to the consumption of reduced fat bakery products: a qualitative approach. CCFRA R&D Report 78

McEwan, J.A. and Sharp, T.M. (1999) Barriers to the consumption of reduced fat bakery products: final report. CCFRA R&D Report 85

Marshall, R.T., Goff, H.D. and Hartel, R.W. (2003) Ice Cream. 6th Edition. Kluwer Academic

NewFoods (2001) A CD-RoM database of new products purchased in the UK 1999-2000. CCFRA

NewFoods (2000) A CD-RoM database of new products purchased in the UK 1998. CCFRA

NewFoods (1999) A CD-RoM database of new products purchased in the UK 1997. CCFRA

Oelke, E.A., Porter, R.A., Grombacher, A.W. and Addis, P.B. (1997) Wild rice - new interest in an old crop. Cereal Foods World, **42** (4), 234-247

Pateras, I.M.C. (1991) Effects of sucrose replacement by polydextrose on structure development of cakes. PhD Thesis. Oxford Brookes University

Ranken, M.D., Kill, R.C. and Baker, C.G.J. (1997) Food Industries Manual. 24th Edition. Blackie Academic and Professional

Sadler, M.J. (1988) Quorn. Nutrition and Food Science. 112: 9-11

Sasaki, M. (1996) Influence of sodium chloride on the levels of flavour compounds produced by the shoyu yeast. Journal of Agricultural and Food Chemistry, **44** (10): 3273-3275

Saunt, J. (1990) Citrus Varieties of the World. An Illustrated Guide. Sinclair International Ltd

Sharp, T.M. (1994) Development and significance of a novel food. Quorn mycoprotein. Voeding, **55** (1): 28-29

Sharp, T.M. (1999) The technical and economic barriers to the production of reduced fat bakery products. CCFRA R&D Report 86

Stadelman, W.J. and Cotterill, O.J. (1973) Egg Science and Technology. AVI Publishing

Taylor, S.L. and Hefle, S.L. (2001) Food allergies and other food sensitivities. Food Technology, **55** (9): 68-83

Thybo, A.K., Molgaard, J.P. and Kidmose, U. (2002) Effect of organic growing conditions on quality of cooked potatoes. Journal of the Science of Food and Agriculture, **82** (1): 12-18

Torres, R.L., Gonzalez, R.J., Sanchez, H.D., Osella, C.A., and de la Torre, M.A.G. (1999) Performance of rice varieties in making bread without gluten. Archivos Latinoamericanos de Nutricion, **49** (2): 162-165

Troccoli, A., Borrelli, G,M., de Vita, P., Fares, C. and di Fonzo, N. (2000) Durum wheat quality: a multidisciplinary concept. Journal of Cereal Science, **32** (2): 99-113

University of Saskatchewan (1999) Production of legume pasta products by a high temperature extrusion process. United States Patent 5,989,620

USDA (1989) United States Standards for Rice. US Department of Agriculture

Wilkinson, H.C. and Champagne, E.T. (2004) Value-added rice products in today's market. Cereal Foods World, 49 (3): 134-138

Wilson, D. (2001) Marketing mycoprotein. The Quorn foods story. Food Technology, **55** (7): 48,50

ABOUT CCFRA

The Campden & Chorleywood Food Research Association (CCFRA) is the largest membership-based food and drink research centre in the world. It provides wide-ranging scientific, technical and information services to companies right across the food production chain - from growers and producers, through processors and manufacturers to retailers and caterers. In addition to its 1700 members (drawn from over 60 different countries), CCFRA serves non-member companies, industrial consortia, UK government departments, levy boards and the European Union.

The services provided range from field trials and evaluation of raw materials through product and process development to consumer and market research. There is significant emphasis on food safety (e.g. through HACCP, hygiene and prevention of contamination), food quality, food processing, food analysis (chemical, microbiological and sensory), factory and laboratory auditing, training, publishing and information provision. To find out more, visit the CCFRA website at *www.campden.co.uk*